Presented By:

Notice to Readers

As this book was going to press, foreign threats of retribution against the pilots of RPA (remotely piloted aircraft) operating in the Mideast were made known to us. In response, we created pseudonyms for some of the characters appearing here, in order to protect their identities. In the rush to get this vital search-and-replace work done, a handful of typographical errors were introduced to the text. Protecting the men and women of our armed forces was our intent, and we deeply regret the error brought on by the urgency of the action. While we hope these errors don't interfere with your understanding of the material, we apologize for any momentary confusion it may cause. If you would like a replacement copy in which the typographical errors are corrected, please send proof of purchase along with your mailing address to:

Zenith Press
Quayside Publishing Group
400 First Avenue North
Minneapolis, MN 55401

Predator

Predator

The Remote-Control Air War
over Iraq and Afghanistan:
A Pilot's Story

Lt. Col. Matt J. Martin
with
Charles W. Sasser

ZENITH PRESS

First published in 2010 by Zenith Press, an imprint of MBI Publishing Company, 400 First Avenue North, Suite 300, Minneapolis, MN 55401 USA

Zenith Press titles are also available at discounts in bulk quantity for industrial or sales-promotional use. For details write to Special Sales Manager at MBI Publishing Company, 400 First Avenue North, Suite 300, Minneapolis, MN 55401 USA.

To find out more about our books, join us online at www.zenithpress.com.

ISBN 13: 978-0-7603-3896-4

Designer: Helena Shimizu
Cover designer: Rob Johnson, Johnson Design, Inc.

Library of Congress Cataloging-in-Publication Data

Martin, Matt J., 1972–
  Predator : the remote-control air war over Iraq and Afghanistan : a pilot's story / Matt J. Martin with Charles W. Sasser.
    p. cm.
  ISBN 978-0-7603-3896-4 (hb w/ jkt)
  1. Martin, Matt J., 1972- 2. Iraq War, 2003---Personal narratives, American. 3. Iraq War, 2003---Aerial operations, American. 4. Afghan War, 2001---Personal narratives, American. 5. Afghan War, 2001---Aerial operations, American. 6. Predator (Drone aircraft) 7. Drone aircraft pilots--United States--Biography. 8. United States. Air Force--Officers--Biography. I. Sasser, Charles W. II. Title.
  DS79.766.M37A3 2010
  956.7044'348--dc22
                                    2010021367

On the cover: *USAF photos*
On the back cover: A U.S. Air Force Predator prepares for landing at Balad Air Base, Iraq, June 9, 2007, after a fifteen-hour combat mission in support of Operation Iraqi Freedom. *USAF photo by Tech. Sgt. Beth Holliker*

Printed in the United States of America

To the men and women of the United States Air Force:
unblinking, persistent, and always on target

Contents

PART III: BALAD AIR BASE, IRAQ

PART IV: NELLIS AIR FORCE BASE, NEVADA

Author's Note

THIS IS A PERSONAL narrative of remotely piloted aircraft (RPA) combat in Iraq and Afghanistan. In it I have endeavored to render the truth as accurately and vividly as possible. This book is an expanded version of notes I kept while actively flying MQ-1 combat missions over Iraq and Afghanistan.

While this is my story, it is also the story of others who played major or minor roles in the events narrated. Actual names are used throughout except in those instances where names could not be recalled or where public identification and exposure would serve no useful purpose and could prove uncomfortable to persons involved.

Dialogue and events are reported to the best of my recollection; some scenes and dialogue have, by necessity, been re-created. Where these occur, I have tried to match personalities with the situation and action while maintaining factual content. While the content is accurate, I cannot be certain every quote is entirely accurate word for word as my interpretation of events may not be exactly the same as someone else's. Time has a tendency to erode memory in some areas and selectively enhance it (or exaggerate it) in others. Where errors in recollection occur, the author accepts full responsibility and asks to be forgiven.

I must also emphasize that although I am an active-duty Air Force officer, the accounts in this book and the occasional opinions I express are my own personal views and do not necessarily represent the views of the U.S. Air Force, the Department of Defense, or the United States government.

I and my coauthor would like to acknowledge the following publications, in no particular order of importance, that have added to the writing and research of this book: *A History of Warfare*, by John Keegan (Alfred A. Knopf, 1994); *From Crossbow to H-Bomb*, by Bernard and Fawn M. Brodie (Indiana University Press, 1973); *The Future of War*, by George and Meredith Friedman (St. Martin's Press, 1996); "Robots at War: The New Battlefield," by P. W. Singer (*Wilson Bobbyly*, Autumn 2008); "The Future of Western War," by Victor Davis Hanson (*Imprimis*, November 2009); "Make Robots Not War," by Erik Baard (New York School of Continuing and Professional Studies, September 9, 2003); "The Army's 'Organic' Unmanned Aircraft Systems," by Major Rangler A. Burdine, USAF (*Air & Space Power Journal*, Summer 2009); "Election Workers Kidnapped in Afghanistan," (AP, Oct–March 2004–05); "Saddam Hussein Defiant in Court," (CNN.com, October 20, 2005); "Journalist Jill Carroll Freed by Her Captors in Baghdad," by Jonathan Finer (*Washington Post*, March 31, 2006); "Hostage Video Ignites Wide Call to Free Carroll," by Peter Grier (*Christian Science Monitor*, January 7, 2006); "What Is the Difference Between Sunni and Shiite Muslims—and Why Does It Matter?" by HNN Staff (HNN.US, December 18, 2006).

Finally, I would like to apologize to anyone who has been omitted, neglected, or slighted in the preparation of this book. While some interpretational mistakes are bound to have occurred, I am certain that the content of this book is true to the spirit and reality of the pilots and aircrews who have served—and are still serving—with the U.S. Air Force in the unmanned aircraft programs. To that end, I am confident I have neglected no one.

—Lt. Col. Matt J. Martin

Prologue

FROM TEN THOUSAND FEET in the sky I peered down upon a large multiwinged building, a technical college taken over by insurgents in the heart of Baghdad. It was after midnight. Streets were unlighted or poorly lighted. Perfect conditions for cockroaches and other vermin to venture out of the gutters. Using an infrared sensor to register heat signatures, I picked out machine-gun and rocket-propelled-grenade (RPG) fire coming from top windows of the college, a blink-blink-blink of muzzle flashes that pinned down a squad of U.S. Army mechanized infantry on the wrong side of the Euphrates River.

I carried a pair of Hellfire missiles beneath my wings, but my task was not to engage the enemy directly. Instead, I was to coordinate with and mark targets for an AC-130U "Spooky" busting its balls to reach the scene. Early versions of the big gunship in Vietnam had been aptly named "Spectre" or "Puff the Magic Dragon." It was the ultimate close air support weapon, an awesome instrument of war equipped with electric Gatling guns that fired six thousand rounds a minute, a 40mm armor-piercing machine gun, and a 105mm Howitzer cannon. Its onboard computers allowed the aircraft's fire control officer (FCO—pronounced "fo-co") to register on my infrared marker and strike with pinpoint accuracy. This was technology and teamwork in action.

As soon as the gunship reported on-site in the night below me, my sensor operator and I began to "sparkle" targets with our infrared (IR) marking laser, lighting them up for IR sensors to detect. The Spooky opened fire with the sound of skies ripping apart on doomsday. Like Armageddon or something. Every fifth round a tracer, it burned ordnance so fiercely that it produced spectacular red cones of fire reaching from air to ground. Death from above. Poor bastards down there in the windows never knew what hit them.

Hostile incoming suppressed, the army ground commander came on the radio and thanked me and the AC-130 crew profusely. Then he saddled up his squad and proceeded on his way. A bit awed by the encounter and its swift resolution, I stood up to stretch and regain my bearings in the "cockpit" of my aircraft. This was my first real-world combat mission following initial pilot training in this particular aircraft. My first ten minutes at the controls of the MQ-1, otherwise aptly known as Predator, and I had already been in on a kill.

Then I remembered that Trish had asked me to pick up a gallon of milk on the way home.

You see, I wasn't in Iraq. Not yet. I was at Nellis Air Force Base, in Nevada, 7,500 miles from Baghdad, flying an unmanned aircraft system (UAS) from a ground control station (GCS). The MQ-1 is about the size of a Cessna 155, except for a longer wingspan. In its Hellfires it carried a big stinger, it could fly at altitudes of twenty-five thousand feet above sea level, and it could remain in the air on surveillance or combat patrol for up to twenty-four hours. The "Nintendo generation," a popular term that caused me to bristle, had taken to the battlefields of Iraq and Afghanistan, where RPAs like the Predator were hindering America's enemies and saving American lives.

Although Leonardo da Vinci was among the first to raise the specter of using flying machines for war, I doubt he could have conceived a time when a pilot could fight a war long distance: commute to work in rush-hour traffic, slip into a seat in front of a bank of computers, "fly" a warplane to shoot missiles at an enemy thousands of miles away, and then pick up the kids from school or a gallon of milk at the grocery store on his way home for dinner.

Now thirty-four years old and a major in the U.S. Air Force, I was a private civil pilot who trained initially as an air force flight navigator before reverting to RPAs in 2005. An RPA or a UAS is popularly known as an unmanned aerial vehicle (UAV). During my years as a "virtual top gun" on two war fronts, I would fly hundreds of missions—and supervise thousands more—in a new kind of combat that, until recently, was largely classified secret. Many aspects of it are still classified. I and other pilots like me would engage in virtually every facet of the Global War on Terror, such as tracking Osama bin Laden, helping rescue hostages, raiding "safe houses" for top al Qaeda leaders, targeting terrorists, and fighting along with the marines.

Sometimes I felt like God hurling thunderbolts from afar.

PART I

Nellis Air Force Base, Nevada

CHAPTER 1

Tonight We Fly

MARCH 26, 2003. SNOW dusted Harir Airfield in northern Iraq during the day, melting into greasy mud and slush by nightfall. In a midnight as dark as the back side of the moon, and a landscape seemingly just as desolate, a flight of seventeen American C-17 Globemaster transports packed with a thousand paratroopers descended from thirty thousand feet toward the airstrip. This would be the first wartime drop for the 173rd Airborne Brigade since Operation Junction City near Katum, South Vietnam, thirty-five years ago, as well as the largest combat parachute jump since the invasion of Panama in 1989.

The jump became necessary only after Turkey balked at allowing sixty-two thousand U.S. troops of the 4th Infantry Division to launch across its southern border. The paratroopers were tasked with securing the rich oil fields around Mosul and Kirkuk and with opening the first northern ground front against Saddam Hussein's regime.

In staggered formation, the fleet roared in at six hundred feet above ground level (AGL) with big engines churning like a tsunami. Jump doors opened. Black wind howled past. Troopers had one minute to catch the wind before the planes passed over the drop zone. There was no need for reserve chutes; if a main failed to open, the unfortunate trooper would smash onto the drop zone before the reserve could deploy anyhow. *Blood an' guts on the risers—and I ain't gonna jump no more* . . . words from an old airborne cadence.

The mighty ships deposited their spores in the air and shrieked on, climbing. A single flashing light from the control tower blinked, blinked, blinked—maybe three times as nearly a thousand parachutes whispered in the dark before thudding their riders to earth. Thirty-five soldiers failed to make it out the doors in time and left with the planes. Twenty were injured on landing, six seriously enough to be evacuated. The force encountered no hostile fire. That came later when Mosul and Kirkuk disintegrated into orgies of looting, arson, and score settling.

Five days before the 173rd launched its northern ground offensive, I arrived on the Greek island of Crete in the Mediterranean with a deployment of three aircraft, six flight crews, and enough support to help open the northern air front with twenty-four-hour, seven-day-a-week operations. By the time I linked up with it, the 55th Wing detachment, redesignated in-theater as the 193rd Expeditionary Reconnaissance Squadron, had been on rotating sustained deployments to the Middle East for over twelve years, snooping and sniffing to help keep the skies over Iraq safe for U.S. and coalition forces. Our defined mission was "to provide vital real-time battle management information to mission planners, commanders, and war fighters."

Each of our four-engine, jet-powered RC-135 Rivet Joint airplanes (a militarized version of the prototype airframe that also led to the commercial Boeing 707) came equipped with an array of antennae and an onboard sensor suite that allowed mission crews to detect, identify, and geolocate signals throughout the electromagnetic spectrum. That meant we could spy on anything electronic in the airwaves within an operations area and listen in on just about everybody. Within seconds after we locked in on a radar or communications signal, we could notify bombers and fighters and have joint direct attack munitions (JDAMs) or rockets screaming down the enemy's throat. It didn't take enemy surface-to-air-missile (SAM) operators long to learn that the sky was listening and watching every time they turned on their radars—and that a missile would be straight on its way toward them. It got to the point where they fired their SAMs unguided in a general direction and prayed for the best. "Allah-guided."

Not knowing what to expect when the northern air offensive kicked off, RC-135 crews were all sucking wind for the first few sorties. Aboard each flight were two pilots, a navigator, and about twenty electronic techs to run the sensors. Cautious of SAMs and MiG interceptors, we flew only night missions as much as possible, each flight anywhere from twelve to eighteen hours long.

An RC-135 is a big, heavy aircraft with a lot of momentum. No 9-G turns for us. Even an expedited 180-degree change of course took a minute and a half. It was a hell of a big target for a SAM and a sitting duck for the Iraqi MiG-25 with its overtake speed of about 1,500 knots.

Although we soon had most of the SAMs intimidated, northern Iraq could still be pretty hairy. It was mountainous country, with peaks up to fourteen thousand feet. The Iraqis took to moving their SAMs every few hours to make it more difficult for us to locate and bomb them. At such elevations, even small SAMs posed a threat and increased our pucker factor fivefold.

Fortunately, the Iraqis had yet to launch a single MiG, at least none that I was aware of. Fearing a repeat of mass fighter defections like those that occurred during the 1991 Gulf War, Saddam ordered bulldozers to destroy runways and taxiways, leaving his warplanes isolated in their hangars. Apparently, he saw his own people as a bigger threat to him than the mightiest army on earth.

With U.S. air supremacy over the theater and with most of the SAMs knocked out in the beginning of "shock and awe," paratroopers of the 173rd on the ground were experiencing more of the war in their first actions than I was apt to ever see. Not that I and the other crews of the 193rd weren't kept busy or weren't putting in long days and nights in "combat." By the time a crew attended mission briefing, conducted preflight, flew the sortie, returned to base, shut down the aircraft, and attended post-flight maintenance and mission debriefs, we had put in a twenty-hour workday.

Tech operators in the aircraft's belly didn't actually start to work until we were on-station, about three hours into the flight. They could nap, read, or talk. Even the two pilots relieved each other at the controls to reduce stress. I enjoyed no such luxuries. As the only navigator aboard, I had to stay awake and alert from takeoff to touchdown.

The plane's sophisticated navigation used an initial gyro system, a Doppler radar, a global positioning system (GPS), and an auto-star tracker that shot celestial headings even during the day, supplemented by air and pitot-static data. Computers took all the information through some complicated mathematical process and spat out our position.

However, since the computers were not always *that* accurate, the navigator had to constantly verify the readings using ground-mapping radar. At short and regular intervals, I compared the picture of the ground presented by radar against charts and the position from the computers. If the pictures did not match precisely, I used a plotter, a pencil, and a pair of dividers to determine the magnitude of the error. Just like from the good ol' days when guys like me flew B-17s over Europe.

This work occupied about half my hours in-flight. Managing the mission to place the aircraft in the optimum position to collect intelligence on targets and avoid enemy threats took up the rest of my time. Since my computer controlled the autopilot, I flew most of a mission from my keyboard.

I swigged a lot of coffee. Not that I complained; I could have been breathing sand and sleeping on a dune. While paratroopers of the 173rd were slogging around in mud eating meals, ready-to-eat (MREs) and getting picked at by snipers, I returned to base and bedded down between clean sheets at Souda Bay on Crete, as brief as sleep sometimes seemed, and awoke to a leisurely breakfast miles away from peril. No wonder dogfaces sometimes viewed the high-flying air force with envy.

I often felt more like a spectator at the singular event of my military generation than an actual participant. Nonetheless, I sometimes enjoyed a fine box-seat view of the fireworks.

One afternoon, B-52 bombers dumped JDAMs on the Mosul area while my plane provided electronic coverage. I counted a dozen explosions, great fireballs, and mushroom clouds of smoke rising into the air. Bomber pilots on the command net begged for more targets. One B-52 turned for home base still carrying thirty-eight bombs.

The U.S. military went to superhuman lengths to avoid civilian casualties. The Iraqis took advantage of our restraint and painted warnings on the roofs of strategic installations: *Contains Human Shields*. That worked until the "human shields" got wind that the Americans were coming. Then they headed for the hills, leaving the target open and ready to receive attention.

Enemy electromagnetic activity all but ceased after a few weeks. With few targets to monitor, I brought along a pair of binoculars to conduct a little sightseeing of the ground war. By now we were flying daylight missions. I saw long lines of U.S. tanks and Kurdish trucks, but no enemy activity. We stared down at Saddam Dam bridging the Tigris River. It had been so well defended before the invasion that coalition fighters patrolling the no-fly zone avoided it. Now we were gawking at it like so many tourists.

On the night of April 16, 2003, less than a month after my detachment deployed, my crew became the first to fly an RC-135 over Baghdad. Saddam had intended to make the city his last stand. A shield of SAMs and antiaircraft artillery (AAA) sites, he claimed, made the city impervious to attack. What he hadn't bargained for was his troops turning off their systems, abandoning their weapons, and walking away.

Most of the city, still without power, sprawled across the desert like a jumble of children's building blocks. A full moon allowed us to gaze down upon the cradle of civilization at the confluence of the Tigris and Euphrates rivers. As there were no windows at the back of the plane, members of the crew came forward to the cabin one by one to look at the miracle of Baghdad subdued and under coalition control.

"I can't believe what we're looking at," someone murmured. "In such a short time."

Everyone had expected the war to drag out for months or more and claim thousands of American lives. Instead, the northern air war was considered over. This was my last sortie. The 55th Wing, after three weeks in combat, would begin redeployment to Offutt Air Force Base on April 17.

It wasn't supposed to be this way. Instead of occupying a box seat to watch history unfold, I should have been a fighter pilot taking on my part of the war, sharing the risks and making history. What none of us realized at the time was that Operation Iraqi Freedom was far from over.

It was never that my future in the air force or any other branch of the military could be considered preordained. If anything, it was an anomaly that I ended up there. I was a farm kid. My folks met at a racetrack. Thoroughbreds, not NASCAR. They were both "poniers," riders who led the jockeys and their mounts onto the track, exercised the horses, and, in general, took care of the animals before and after a race. I was all but born on a track. In between races, as it were. The eldest of two sisters and a brother, I traveled with my parents to tracks all over the United States, living in trailer parks.

When I was about twelve, Dad ventured into the hay business, buying a ranch in Indiana to grow and supply fodder to racetracks. The family had grown too big by then for the footloose life. I went into business with Dad and grew, cut, baled, and hauled hay until I left for college. It was hot, back-breaking labor, with sweat stinging my eyes and hay rash making my crotch and arms itch. On July days when an airplane—it didn't matter what kind—would fly over the field, I stopped whatever I was doing, leaned against the baler or the truck, wiped my face with a bandana, and gazed longingly after the airplane until it disappeared.

"Matt the Cat. Come down to earth, Son."

That was my dad's nickname for me. I suppose it had something to do with my restless nature. My mind often wandered into the skies, catching those wonderful airplanes that transported my imagination around the world.

"Yeah, Dad. I was just wondering . . ."

I continued to wonder. I saw *The Right Stuff* twelve times. I yearned to become a fighter pilot.

I graduated from Purdue University in 1994 with a commission as a second lieutenant through the air force ROTC program. To my disappointment, flight training was temporarily closed to volunteers, and I ended up attending missile school. Afterward, I was the officer in charge of an underground nuclear missile silo near Cheyenne, Wyoming, waiting for orders to launch and annihilate the world. The irony—while my dreams soared above the earth, my body was buried underneath the earth.

Since flight training remained closed, I volunteered to be a navigator, figuring I could get what I wanted through the back door if not the front. At least I would be in the air. I trained in T-34s at Pensacola Naval Station and T-43s (the military version of the Boeing 737-200) at Randolph Air Force Base, graduating near the top of my class. At the same time, I acquired a civilian private pilot's license with an instrument flight rule (IFR) rating, hoping it would increase my competitiveness for air force flight school.

Catch-22. The air force changed its rules while I was attending nav school. I now had to complete two and a half years as a navigator in order to fulfill my obligation to the Department of Defense for the training I received. That would make me thirty-one years old and ineligible for flight training; the new cut-off age was twenty-seven. It was a crushing blow. It seemed I would forever be a bridesmaid and never the bride.

When Operation Iraqi Freedom (OIF) kicked off in 2003, I was stationed at Offutt Air Force Base, in Omaha, Nebraska, head-Bobbys of the U.S. Strategic Command, the Air Force Weather Agency, and the 55th Wing of the Air Combat Command. Prior to OIF I had deployed to Prince Sultan Air Base in Saudi Arabia a half-dozen times to patrol the southern no-fly zone over Iraq. I also spent a fair amount of time at other hot spots around the world conducting what we called "sensitive reconnaissance." This included missions over Afghanistan during the opening days of Operation Enduring Freedom (OEF).

To kick off OIF, I deployed to Crete, then redeployed to Offutt after not quite a month in Southwest Asia with the feeling that my career in the air force had so far been rather unremarkable.

I married Trish shortly after redeployment. We had met three years previously in Omaha. She had been a theater major in college and I was involved off-post with an improvisational theater group. She attended one of my shows and came to the after-hours party. As bodice rippers are wont to write, I looked across a crowded room and there she was. Lovely, slender, red haired, with a shy smile and warm blue eyes.

Trish was a woman with a heart big enough to embrace kids, animals, and a would-be air force pilot. My family began calling her Ruby, not only because of her hair color but also because there were already two other Trish Martins in the family, including my middle sister. Trish soon sensed that I was not quite satisfied with the lot I had drawn in the air force. It wasn't exactly that I was discontented, rather that I thought I should be doing more.

The war in Iraq had not ended when special forces dragged Saddam Hussein out of his rat hole. Instead, the United States found itself becoming mired in a sectarian insurgency that pitted Shiite Muslims against Sunnis and elements of both against the United States and the U.S.–backed Iraqi government. Increasingly restless, I often thought of the night the 173rd parachuted into the mud at Harir Airfield. Those guys had contributed up front and personal while I had never even gotten my feet muddy.

"You did your part, Matt," Trish reassured me.

"But it's not enough. Don't you understand?"

I was perusing an air force assignments website when I came across a notation soliciting Predator pilots. All I knew about Predator was that it was a remotely piloted aircraft, an unmanned airplane. Not exactly a fighter, although it was armed. It sounded almost like science fiction. Nonetheless, I recognized an opportunity to fly, if only by proxy, and get back into the war. Not in the mud and blood with paratroopers, but certainly closer than thirty thousand feet in the night sky above them as little more than an observer. And certainly nearer than Offutt Air Force Base.

I met the requirements since I was an experienced navigator with a commercial pilot's license. I called the assignments officer at the personnel center. He eagerly signed me up as a volunteer (the program enjoyed few volunteers in those days).

"If you're accepted," he cautioned, "it means you'll likely deploy to Iraq or Afghanistan sooner or later."

I understood. I was already looking forward to flying combat missions. Little did I realize that the war for me was about to begin in a way I could never have contemplated.

CHAPTER 2

Feel the Plane

I WAS A DELIBERATE man in spite of Dad's "Matt the Cat" nickname. I wasn't the type who, upon hearing about the Confederates firing on Fort Sumter, for example, would have started waving the flag and shouting for action. My volunteering for the Predator program may have appeared whimsical on the surface, but that couldn't have been further from the truth. I had been preparing to fly for the past ten years, almost from the moment I stood in the sun to be commissioned an officer in the U.S. Air Force out of Purdue ROTC.

Trish and I packed up, moved out of our duplex, and drove from Offutt to Creech Air Force Base in Nevada. If you could have cut out a hunk of Iraq's terrain around Baghdad and matched it to anywhere in the United States, that would have been in Nevada. Various shades of desert brown crumpled by the fist of a giant stretched to the horizon, broken only by green along waterways. Creech was in the middle of nowhere, almost as though it contained secrets not open to everyday scrutiny.

An air force lieutenant colonel, who introduced himself as a *real* fighter pilot, met me and two other trainee candidates when, blinking in the bright sun, we arrived on-base for orientation to the Predator Formal Training Unit (FTU). He was a big man with craggy features, a gray burr cut, and piercing, no-nonsense blue eyes. He gave us a tour of the flight line, escorting us to a hangar open at either end in the middle of which sat a strange extraterrestrial-looking gray airplane without a cockpit or windows.

"You might think of it as a big, super-fancy, remote-controlled model airplane," the lieutenant colonel said. "You couldn't be more wrong. Gentlemen, what you are looking at is the future of modern warfare."

Part of my nature—call it anal retentive—was to thoroughly investigate in advance every aspect of any new enterprise upon which I intended to embark. I had therefore researched the Predator completely before arriving in Nevada for training. I knew that it was 27 feet long, had a 48.7-foot wingspan, and was powered by a Rotax 914 four-stroke 115-horsepower rear-mounted turbocharged engine with a two-blade, variable-pitch pusher propeller. It could reach speeds of 135 miles per hour, attain altitudes of 25,000 feet, and stay airborne for up to twenty-four hours without refueling. I thought it mildly curious that you started the engine the way you would a modern outboard boat motor—by pushing a button located in the ground panel on the outside of the plane. Earlier models had used a pull cord. You killed the engine by running onto the taxiway after landing and hitting a "kill" switch in front of one of the wings.

"The only thing better than having a robotic airplane assisting on the battlefield," our guide noted, "is having one that will fight for you."

He showed us mounts underneath the wings designed to carry twin AGM-114 Hellfire missiles.

"It's a big bee, gentlemen, with one hell of a sting."

The pilot "stung" through what he called a multispectral targeting system (MTS) located in a ball under the aircraft's nose. An enlisted "sensor operator" aircrew member in the "cockpit" with the pilot, performing copilot-like duties, fired a laser or an infrared beam from the MTS. The laser beam reflected off the target and pulsed to attract laser seekers in the Hellfire missile. Either that or an onboard computer used the laser beam to calculate location, distance, and other battlefield variables that could be transmitted to manned aircraft or ground forces so they could destroy the target. This was called "lasing the target." "Sparkling the target" was the term used for the infrared (IR) marker.

"The Predator is very unobtrusive, the Hellfires are supersonic, a combination that gives little advance warning of an attack," we were told. "The bastards never know what hits them."

A pilot "flew" the plane through a full-color nose camera that transmitted images to a ground control station (GCS). The MTS also came loaded with a variable-aperture camera, like a traditional TV camera and an infrared camera for low-light and night flying. All the information obtained by the airplane's sensors could be transmitted to other battlefield commanders and stations with a need to know.

I was profoundly impressed. Even visionaries like H. G. Wells, whose short story "Land Ironclads" inspired Winston Churchill to champion the development of the tank, or Jules Verne—*Twenty Thousand Leagues Under the Sea*—could not have conceived of something like this.

The tour continued with the lieutenant colonel leading the way to a thirty-foot-long enclosed trailer parked near the main classroom. This, he explained, was the GCS. In effect, Predator's "cockpit." It contained consoles for the pilot and his sensor operator as well as data exploitation and mission planning consoles. The pilot's station resembled a traditional airplane's cabin or cockpit that used a standard flight stick, rudders, trim tabs, throttle, and other associated controls that communicated commands electronically to the Predator while the pilot looked through the airplane's cameras.

The GCS trailer was designed to be rolled into a C-130 Hercules for transport. The plane itself could be disassembled into six main components and loaded for transport into a container nicknamed "the coffin." A crew of four could reassemble the system on-site and have everything ready to fly within twenty-four hours.

Experimental reconnaissance drones had existed since at least the early 1980s. Both the Central Intelligence Agency (CIA) and the Pentagon had researched them. An early CIA model called "Gnat" sounded like a "lawnmower in the sky." General Atomics Aeronautical Systems developed a prototype Predator with a quiet engine in 1994 at a cost per operating unit of about $3.2 million. The price was now more than $4 million.

Since then, Predator had flown into combat alongside manned warplanes to provide air support for ground forces and furnish intelligence to forces commanders, its battlefield effectiveness tested in Bosnia, Kosovo, Yemen, Afghanistan, and Iraq. Predators first deployed to Europe in 1995, based in Albania. They were pulled out after Serb air-defense gunners shot down two of them.

During the Kosovo air campaign in 1999, Predators collected intelligence, searched out targets, and kept protective cameras trained on Kosovar-Albanian refugees. In September 2000, the CIA and the Pentagon joined in an operation called Afghan Eyes, the purpose of which was to use Predators to locate Osama bin Laden. About that same time, White House security chief Richard A. Clarke suggested that armed drones might eventually be used to target bin Laden.

Predator's role as a strike vehicle began with a series of Nevada range tests in February 2001. A friend of mine named Curt "Hawg" Hawes was one of the test pilots. The aircraft fired three AGM-114 Hellfire antiarmor missiles on February 21, scoring perfect hits on a scrapped stationary tank. Further tests occurred between May 22 and June 7, with equal accuracy. The drone was subsequently designated as an MQ-1B. The M stood for "multirole," the Q referred to an unmanned aircraft system, and the 1 described it as being a first in the series.

After the terrorist attacks on the Twin Towers and the Pentagon in September 2001, Predators armed with Hellfires were deployed to Afghanistan within the week, where the first missions with weapons were flown over Kabul and Kandahar. The Hellfires' most infamous and controversial utilization came through stealthy aerial assassinations. Snipers in the sky.

On February 4, 2002, an armed Predator targeted a convoy of sport utility vehicles and killed a suspected al Qaeda leader. On November 3, the car of an al Qaeda terrorist named Qaed Senyan al-Harethi was speeding along a road in Yemen; the car was occupied by five men suspected of being responsible for the bombing of the American destroyer USS *Cole*. One Hellfire from a CIA Predator flown by a pilot from a French military base in Djibouti took out the vehicle and killed all five terrorists. On at least two other flights, Predator cameras spotted a tall man in white robes at bin Laden's Tarnak Farm outside Kandahar, in Afghanistan. Although the man was likely the notorious terrorist himself, subsequent unclassified reports revealed that the pilots could not obtain clearance to fire him up.

The Predator expanded its usefulness on March 4, 2002, when a U.S. CH-47 Chinook helicopter crashed on top of Takur Ghar Mountain during Operation Anaconda in Afghanistan. A joint Taliban–al Qaeda machine-gun crew pinned down the army Ranger team being transported by the chopper. In its first close air support role, a Predator fired a Hellfire missile into the reinforced enemy machine-gun bunker and relieved pressure on the Rangers.

At the end of December 2002, three months before the onslaught of Operation Iraqi Freedom, Predators armed with AIM-92 Stinger air-to-air missiles flew over Iraq's no-fly zone to "bait" Saddam's MiGs. Most of the time the Predator fled when detected. One of the aircraft, however, fired a Stinger at a MiG -25 instead of trying to escape, the first time in history that a conventional aircraft and a remotely piloted aircraft engaged in air-to-air combat. The RPA lost and was shot down when its Stinger heat seeker become "distracted" by the MiG's missile.

By the time I arrived at Creech Air Force Base in 2004, the war in Iraq was moving into its second year and showed no indications of slowing down. If anything, it was going to get a lot hotter. Congressmen and antiwar protesters were already calling it a "second

Vietnam." Whereas at the kickoff of Operation Iraqi Freedom, older Predators were stripped down and used as decoys to entice Iraqi air defenses to expose themselves by opening fire (we called them "chum birds"), they were now being deployed as combatants.

"Initially," explained our high-ranking guide, "the pilot and the sensor controlled the Predator from a GCS via a C-band line-of-sight data link."

I already knew that. It meant the controllers were in-country flying the drone from its local operating base, to which it was restricted within a hundred-mile radius. My jaw went slack when the light colonel completed his explanation.

"Deployments now use a Ku-band satellite data link for beyond line-of-sight operations. That means— "

Stunned, I finished the thought for him. "You can fly it in Iraq from here at Creech?"

"Isn't that amazing?" He beamed.

"Amazing," I concurred.

And I thought I was going to get closer to the war than thirty-thousand feet overhead in an RC-135.

"Captain Martin! What the fuck's the matter with you? Got your head up your ass?"

"What the fuck, *Mister* Moran!" I shot back. "I got it, okay?"

Flying the blasted thing was all about confidence. I couldn't let him shake mine. The Predator wasn't so difficult once you got it in the air. Takeoffs weren't too bad. Landing it, however, could be a real bitch.

I had a stiff Bobbying crosswind on final approach to landing, the runway in sight through my nose camera. It was a five-thousand-foot stretch of asphalt shimmering in the Nevada sun, running past low-slung open hangars where the training aircraft were kept. This was my third time on approach, having aborted the first two accompanied by Instructor Moran's derisive comments. Negative reinforcement was how he taught. You had to be prepared to catch flak from him at all times. We both knew the statistics. Every one hundred thousand flight hours in a Predator resulted in eight or nine crashes that could have been fatal were the aircraft occupied.

I was determined not to give Moran the satisfaction of seeing me sweat.

"You're *in* that airplane, Captain Martin. *Feel* it."

I suppressed an urge to tell him to shove it.

"I feel it, Mister Moran. I *feel* it."

Steve Moran was always preaching how the best Predator pilots *felt* the airplane, even though they weren't actually in it. He stood over my shoulder in the GCS with one of the sensor students in the seat to my right. He was a solid hunk of civilian in his fifties and a former air force F-4 pilot, a little detail he never let any of the ten student pilots and eleven future sensor operators in my Predator training class forget. Contracted by the air force, he was supposed to be one of our primary instructors for the full three months of training. He was old-school when it came to military education. He should have been wearing a marine drill instructor's hat with a brim as sharp as a Skilsaw blade. His jaw stuck out like a mule's. *Whassa matter with you shitbirds, huh? You are slow, you are sloppy, your breath stinks, and you don't love Jesus. You ain't gonna make it, ladies.*

By now, I had logged hundreds of hours as a private pilot in a variety of civilian airplanes, plus more than a thousand as a military navigator in the RC-135, including combat tours over Iraq. I knew how an airplane was supposed to feel—and sitting in a GCS, for all it *looked* like a flight cockpit, wasn't it.

Conventional airplanes were flown with direct mechanical or hydraulic systems. They were designed to be stable in the air. In a plane properly trimmed, the pilot could move the control stick or yoke into a left or right bank turn, and ascend or descend, and the plane maintained that attitude on its own, continuing the maneuver until the pilot altered it.

The same inertia and acceleration that influenced the airplane also affected the pilot. He felt gusts of wind, turbulence, a change in the aircraft's relative position to the ground. It was "seat of the pants" flying, upon which airmen had depended all the way back to Wilbur and Orville Wright.

The Predator pilot had no such connections to his plane. Controls in the GCS were spring loaded, which meant the control stick and rudders had to be held in place if the airplane were to maintain an attitude. Once the stick was released, it popped back into center position and the airplane returned to straight and level flight. Furthermore, control input went into a computer, which in turn transmitted instructions to another computer in the airplane that ordered Predator's servo to execute the change. That meant a lag between the command and the response, one that became ever more pronounced when using the satellite Ku-band. It took some getting used to.

The Predator flew more like a glider than an airplane, having been designed with very large wings compared to its fuselage size. While glider-like wings were the secret to its long flight duration, they made the aircraft extremely sensitive to wind and turbulence. That was particularly relevant when operating close to the ground, such as during landings and takeoffs. The sun was always doing nasty things to the air on the desert runway at Creech.

What I *felt* mostly was Moran behind me in the GCS, his hand clawed over my shoulder as though prepared to jump in and grab the stick if necessary. Unlike regular flight training, there were no dual controls for the instructor. I was on my own. Mounting tension burned my eyes with sweat, but I knew I could do this. I was a Gen-X'er, a generation that grew up with computers. And I knew how to fly small airplanes.

Since the nose flight camera in the aircraft was fixed, there was no way to shift my observation off the runway. I had a thirty-degree field of view, compared to the fifty degrees of the naked human eye. It was like trying to fly while looking through a soda straw. Like riding a roller coaster without being able to turn your head or look up or down.

I crabbed the craft into the crosswind. A gust caught it. I lost sight of the runway. My screen filled with white hot sky. I thought Moran was going to shit his pants.

"I got it," I insisted. "I got it."

I eased down the nose. I was off to the left of the runway. I compensated. I didn't feel the ground rush, simply saw the landing strip on my computer screen. I pulled back slightly, reduced power. The last thing I needed was to land on my nose wheel and porpoise down the runway. That meant I had washed out and would be on my way back to Offutt.

I could see the nose of my plane through the GCS screen with the runway unfolding in front of it. I suddenly crouched forward to look over the nose to get a better view, a futile gesture when flying an RPA. I heard Moran chuckle.

"*Now* you're feeling it," he said.

It was one of the smoothest landings I ever made, either in a real airplane or the drone. I completed the rollout and cut power. I taxied off the runway and glimpsed the ground crew on my screen running out to hit the kill switch and push the Predator to parking. I relaxed and turned to Moran. I couldn't contain a smug grin.

"Not too bad, shithead," he growled.

That was the nearest he ever came to a compliment.

The weeks passed quickly. The class progressed through the various training phases—in-flight emergencies, use of sensors, turnover procedures to relief crews, launch and recovery element (LRE) methods, surveillance techniques. Finally, live fire on the range with Hellfire missiles against fixed targets such as hastily erected composite building structures (CBS buildings) and junk cars. My first shot, I "pickled" and watched the pixilation and sudden exhilarating brightening on my screen as the missile impacted right on-target, a sequence that would become all too familiar over the next few years.

Not all the students were able to overlook Moran's ranting and frequent tantrums. Lieutenant Richard Williams, a former B-52 pilot from Barksdale Air Force Base, and I became friends. Chuck was a capable pilot, but sometimes he let Moran rattle him.

"He's trying to wash me out. He wants to flunk me."

"That's not it," I argued. "He thinks he has to put on the pressure to see how we respond."

"Do you think he really was a marine Full Metal Jacket DI in another life?"

During the twelve weeks of class, students scraped up three aircraft tailboards at a replacement cost of ten grand each; crashed one aircraft, destroying it, to the tune of about four million; burned up thousands of gallons of aviation fuel; and shot off eight Hellfire missiles at sixty-eight thousand dollars per. I figured Uncle Sam spent a cool Bobby million dollars training me to be a Predator pilot.

My friend Lieutenant Williams almost washed out the last month. With Moran looking over his shoulder, he was returning from a practice surveillance flight, coming in on short final through ground turbulence. As usual, Moran was shouting and getting red in the face.

"You're slow!" he raved. "Get down into the flare! You're not lined up with the runway."

Suddenly, Moran really lost his temper. He reached over Chuck's shoulder and grabbed the stick. The screen went crazy for a second or two as the Predator skidded, skipped, and bounced down the airstrip before somersaulting and ending up in a big hunk of scrap kevlar.

There was an investigation and Moran was fired. The instructor who succeeded him had a quieter disposition. He probably *hadn't* been a marine DI. After undergoing additional correction training, Chuck finished with the rest of the class. He couldn't contain his elation. A black pilot who drew inspiration from the Tuskegee Airmen, he strode up to Trish and me following graduation ceremonies, hand extended.

"Kick the tires and light the fires!" he exulted. "We're off to war. It's the real thing from now on."

I hesitated. I looked out toward the Predator hangars and the stationary GCS trailer, safe here in Nevada, from which we would presumably be operating armed aircraft in battle over Iraq and Afghanistan.

"The next generation goes to war," I said with more sarcasm than I intended.

CHAPTER 3

First Mission

NELLIS AIR FORCE BASE, Nevada. My first real-war Predator mission. I was a bit nervous, just enough to keep my senses honed and alert as the off-going pilot briefed me on the situation before we swapped out and I took over for my shift. The airplane was ten thousand feet above Baghdad covering the withdrawal of an army mechanized patrol. It was after nightfall in Iraq. The infrared screen showed several Bradley fighting vehicles crawling through the slums in the direction of the Euphrates River.

Predator was the perfect air-cover weapon. It could remain in the air for a full day and night hunting or staring at a target. When it came time for it to go home and refuel, another Predator took its place. It was the same with the operators. We had unlimited patience. We were always present over the war front, watching, waiting.

I took my seat as I had done many times during training at Creech, adjusted it to my liking, put on my headset, plugged it in, and conducted a normal changeover checklist, allowing me to confirm the status of my airplane that was already in the air 7,500 miles away. I scanned my instruments—airspeed, altimeter, fuel and oil levels, power settings, system voltage, engine power, and servo temperatures. I tested my control stick, throttle, flap lever, rudder pedals, ailerons, and propeller revolutions per minute. I confirmed that I had the correct maps, imagery, chatrooms, and concept charts pulled up for reference. Everything was good to go.

It was like a sci-fi novel. Here I sat "flying" in Nevada while my warplane orbited over Baghdad. An LRE on the ground at Balad Air Base, a protected airstrip in Iraq, had launched the drone and would land it because of the satellite signal delay between my GCS and the airplane. All I had to do was fly the aircraft once it was safely airborne.

I managed this over such a long distance through a complex system of fiber-optic cables and satellite dishes. Phone lines carried flight commands from the GCS across the nation and under the Atlantic to a satellite dish in Europe. The satellite dish bounced the signal to another small dish on the Predator. Over the same route, the aircraft in turn sent back one telemetry and two video streams, which were then displayed on my various GCS screens. The string of computer nodes caused a two-second lag between my input of a command and the aircraft's response, thus the need for the in-country LRE.

The GCS featured four screens in front of both the pilot and the sensor operator, and a ninth screen mounted between them. The tracker, the top screen in front of me, displayed a map of Baghdad and an icon to indicate the location of my plane. Below that the heads-up display (HUD) showed video from the aircraft targeting pod, where I watched the Bradleys wending their way through Baghdad. Below that, by the flight controls, were two heads-down displays (HDD 1 and 2), which showed engine and fuel data and other aircraft information. The mission computer, between me and my sensor operator, was split down the middle: the right half was a map showing the positions of all U.S. air and ground forces in the vicinity as represented by appropriate icons, the left half was divided into several "chatrooms" like those on the ordinary Internet. Except these chatrooms functioned through a separate classified military Internet.

Chatrooms were how I primarily communicated with supported units and air traffic control in Iraq. At the moment, an army intelligence specialist supporting the mech patrol was using a computer linked into the worldwide military net to type his messages. I typed messages back to him. He also used a second computer to receive my video signal from the aircraft. This was definitely not "your father's war."

Everything in Baghdad seemed to be going along fine for the mech patrol until it encountered hostile fire coming from the top windows of the technical college. The radio came alive between me in my GCS and the soldiers under fire and between me and the AC-130 dispatched to suppress the fire. Only ten minutes had elapsed of my first real mission and I was already right in the middle of an action. I felt a tightness in my chest. My breathing shallowed.

Near my RPA icon on the map screen appeared another icon representing the AC-130. On the HUD, my infrared picked up the flicker of machine-gun fire erupting from several windows at the college. The AC-130 asked me to "sparkle" the target with an infrared laser to mark it. He would do the rest—which he accomplished in fantastic style with a cone of Gatling fire. The Bradleys thanked us and proceeded on their way to the Euphrates. Repress and move on.

The suddenness of action played out long distance on computer screens left me feeling a bit stunned. A surreal experience. Almost like playing the computer game Civilization, in which you direct units and armies in battle. Except with real consequences.

I also felt electrified, adrenalized. *My* team had won. We had shot the technical college full of holes, destroying large portions of it and killing only God knew how many people. It would take some time for the reality of what happened so far away to sink in, for "real" to become *real*. As Gen. Robert E. Lee so famously quoted during the Civil War: "It is good we find war so horrible, or else we would become fond of it."

I had yet to realize the horror.

***

Soldiers liked to say it was a good war if everything was quiet within their own little corner of it, no matter that hell may be erupting a short distance away. Whereas most ground units such as the 173rd Airborne that parachuted onto Harir Airfield became more or less confined to a single area of operations (AO), airpower has no such "little corner." Through the miracles of modern technology, I could fly anywhere in the world from the Predator Operations Center (POC) at Nellis, even opposite points on the globe on the same day, provided there was a Predator and an LRE available on-site. That meant I participated in wars in both Iraq and Afghanistan.

The concept that we could topple the Taliban with a minimum number of troops, establish a new government, then leave as quickly as we came turned out to be a significant miscalculation on the part of our senior leaders. What was supposed to be a short conflict had now dragged on for three years with no end in sight.

I used to think Afghanistan was not such a big country. An RC-135 flying at five hundred miles per hour could have lapped the entire nation in about two hours. When the war began shortly after 9-11 and the Twin Towers, the skies over Afghanistan were so full of air traffic that airplanes running into one another was a greater threat than anything the Taliban could muster. Now, three years later, about the only craft flying were a couple of A-10s to support ground contact and the occasional B-1B bomber with its fourscore 500-pound bombs. And, of course, Predator skulking about to keep an eye on what might be going on down there.

In Iraq, Predator operators piloting from Nevada and handling LRE in-country were air force. It was different in Afghanistan. Pilots were mostly air force like me, but guys on the ground running the operations and handling LRE seemed a rather shadowy bunch. No one ever knew their names. In chatrooms, they were referred to only as "Op-4." I assumed, correctly or incorrectly, that they were CIA or CIA-affiliated former marines and army special forces. For all I knew, they might be flying their own Predators against the Taliban by line of sight. Either way, I did not have a need to know.

Predators in Afghanistan were unarmed, at least those we flew from Nellis. They carried no missiles, only cameras. That would soon change, however. But until it did, I figured sooner or later we were bound to scare up an important target—say, Osama bin Laden—and let him get away before we could get firepower on the scene.

The sun was shining brightly in Nevada when I entered the GCS trailer for one of my first reconnaissance missions against the Taliban. I switched out with the off-going pilot and entered my alternate world and persona. I noticed on the HUD that night had fallen over Afghanistan and the infrared cameras were up and going. Flying Operation Enduring Freedom could be almost as exciting as watching paint dry. Tonight was going to be a caffeine overload night.

"Keep the pots brewing," I joked with the POC staff.

Afghanistan, unlike Iraq with its large population centers, was mostly rural. Therefore, the search for terrorists centered on farms and mountains over which we could fly for hours without spotting anything more interesting than some local beating his camel (or his wife if he was Taliban).

During my preshift briefing in Ops from the mission commander (MCC—to distinguish it from MC, a mission coordinator) and again from the relieved pilots, I learned that we had an objective for the night.

"We picked up a hotspot right after dark," the MCC explained. "The intel officers think it might be an underground bunker like we've seen before. Maybe a weapons cache."

Infrared recognized differences between objects. On my screen I saw a darkened barnyard—a shed of sorts for livestock, a little mud house with no lights, a couple of Genghis Khan ponies, and a camel wandering around munching on whatever surviving forage it could find. The "hotspot" lay between the barnyard and the house with a pathway leading to it from both directions. I couldn't tell what it was. Visibility through infrared was somewhat limited. Our supported unit in Afghanistan wanted me to keep an eye on it until daybreak allowed a positive identification. Until then, invisible in the night sky, I would continue to circle and stare at it, waiting for something to happen.

I was already starting to refer to the Predator and myself as "I," even though the airplane was thousands of miles away.

Hours passed. Nothing happened below other than horses and camels sometimes switching places or intermingling. Once, a Taliban camel landed a few swift kicks into the side of a Mongolian horse, which dodged away. I yawned.

Once the sun rose, bathing the rugged terrain below in a soft pink light, I asked my sensor operator to switch to daylight cameras. The telephoto lenses brought out the full picture. A guy with his head wrapped came out of the house to stretch and yawn, completely unaware that he was being watched. He kicked half-heartedly at a yellow dog, and the cur ran away down the cart path that led to the "hotspot" we had fixated on all night.

In the full disclosure of daybreak loomed a large mound. My experiences of having grown up on a hay farm told me what it was right away. I burst into laughter. The feature that we had stared at for hours, that had preoccupied the most sophisticated reconnaissance apparatus on earth and baffled the world's finest intelligence analysts, was a pile of barnyard manure.

Flying over Fallujah in Iraq's Al Anbar Province in August 2004, I peered down upon a city of nearly a half-million inhabitants that had become a sort of Islamist ministate with Sharia law enforced by al Qaeda and their sympathizers. Rampant interfaction fighting produced gangs of thugs, terrorists, and would-be martyrs prowling the streets to beat up or execute shop owners, barbers, or anyone else who dared to so much as think "Western style," much less practice it. Most music was banned. Women returned to covering their faces in public. Shouting, threatening gunmen roared through the city in "technicals"—pickup trucks with .50-caliber machine guns mounted in the beds. It was a lot like Beirut must have been in the 1980s or Afghanistan in the 1990s.

Fallujah was in the middle of a turbulent and violent recent history. Currently, it was a no-go zone for U.S. military ground forces. Marines enclosed the city in a cordon of steel, nobody in or out without proper authorization, while tactical U.S. warplanes continued to target insurgency "safe houses" and weapons cache sites within.

The occupation of Fallujah began in April 2003, one month following the invasion of Iraq. Although a majority of the residents were Sunni and therefore had supported Saddam Hussein, the city was one of the most peaceful regions in the country—for about one week.

The U.S. 82nd Airborne entered Fallujah on April 23 without resistance and took up positions in three primary locations in the city—at the vacated Baath Party headBobbys, at a Baath Party resort just outside city limits, and in a local center-city schoolhouse. It took only five days for resentment to build. On April 28, crowds protesting the U.S. military presence descended upon the school, where elements of the 82nd were headBobbyed. A confrontation ensued. Shots rang out. Paratroopers said insurgents in the crowd fired on them first. American soldiers returned fire, killing seventeen Iraqis and wounding more than seventy.

Summer approached. The city grew increasingly unsettled and hostile toward the occupation. In an effort to cool things down, U.S. troops turned the "law and order" mission over to Iraqi police and the Iraqi Civil Defense Corps and moved out of Fallujah to positions on the outskirts. That didn't work. On June 30, a huge explosion at one of the two hundred mosques in the city killed an imam and eight others. Naturally, it was blamed on the United States, which had avoided violating the mosques so conscientiously that insurgency leaders felt safe in stockpiling their weapons in the shrines while they plotted against coalition forces. American commanders suspected that the explosion was either the result of a deliberate bombing by insurgents to incite the population or an accident involving a bomb training class.

In its wake, urban guerrilla fighters launched a number of martyr attacks on police stations in the city, killing themselves and at least twenty policemen. Even though the Koran specifically forbade suicide, local imams were apparently successful in convincing dozens of Allah's children to throw their lives away. No imam was identified among the martyred dead.

The city remained restless and unsettled for the rest of the year, violence rapidly increasing after New Year's 2004. On March 31, 2004, a gang ambushed a convoy guarded by private military contractors employed by Blackwater USA. The convoy of American civilians working with Eurest Support Services was delivering much-needed food to hard-hit civilians in the city. Four of the contractors were pulled from their cars and then beaten and set afire while they were still alive. Their burnt corpses were dragged through the streets by jubilant mobs in macabre scenes reminiscent of Mogadishu. The corpses were finally hung in display off a bridge spanning the Euphrates River.

Worldwide outcries demanded a response to the atrocity. U.S. Marines responded with Operation Vigilant Resolve and laid siege to the city. Iraqi National Guard troops immediately discarded their uniforms and deserted. The United States suffered forty deaths and killed over three hundred Iraqis during the siege and subsequent raids to capture perpetrators and others involved in insurgent and terrorist activities.

The U.S. military declared a unilateral truce on April 10 and pulled back to the outskirts to allow humanitarian supplies to enter. An Iraqi mediation team arrived in an attempt to set up negotiations for a lasting truce. Insurgents responded to the "ceasefire" by capturing two U.S. soldiers, seven U.S. military contract employees, and more than fifty civilians and humanitarian workers. Some were later released, but most were brutally executed, some even beheaded and the atrocities filmed for distribution through the Internet.

Still striving for peace, marines were persuaded to hand over control of the city once again, this time to a former Iraqi general and his thousand-man Fallujah Brigade, who promised to restore order. The mosques and insurgents proclaimed victory. Celebratory banners appeared all over the city, waving above shouting crowds. Armed fighters paraded about in their technicals. Weapons were found hidden in United Nations (UN) humanitarian supply trucks attempting to break roadblocks.

The Fallujah Brigade was promptly and predictably marginalized and became merely another faction in the strange internecine warfare that boiled the city in its own juices. The big cheese in town was one Abu Musab al-Zarqawi, reportedly third in command to Osama bin Laden and leader of al Qaeda in Iraq. The United States continued sporadic, intelligence-based air strikes against safe houses used by al-Zarqawi and his al Qaeda henchmen while threatening to return to Fallujah in force to restore order.

That was the situation in August 2004 as I peered down from the sky into battle-torn city streets. My job was to find targets, al-Zarqawi if I were lucky. I was a patient, silent hunter. I was armed.

CHAPTER 4

The Green Toyota

HUNTING TERRORISTS AND INSURGENT fighters required indefatigable patience. The business of intelligence gathering was long, slow, tedious, unglamorous work only occasionally punctuated by intrigue and excitement. That was true whether in the clandestine "in the enemy's camp" business or flying a multirole RPA. We could have sent troops busting through Fallujah rooting out the cancer house by house, except that would have been extremely dangerous and would have taken a lot of troops. Fallujah was a big city. Instead, we used spies and informants on the ground and Predators in the air (along with other airborne assets) to locate and track bad guys until we could take them out. With prejudice.

The process of locating a suspected terrorist or an insurgency leader more or less followed a set pattern. It often began with some informants sneaking up to one of the marine outposts that still ringed the city. Most of the time he came in the middle of the night to keep from being seen or recognized by his neighbors or fellow Iraqis. If al Qaeda learned of his perfidy, he became the next dead man hung from a light post or left beheaded in the middle of the street.

His motive for informing, therefore, had to be strong enough to overcome his fear. Frequently, he sought revenge for some slight or perceived wrong committed against him, which made the information he supplied less reliable than if he were prompted by mere greed. Greed produced the best information.

"I know of a house where they have many weapons. You pay me, I'll tell you where it is . . ."

The marines would then nominate targets via intel channels, which would often be translated into a request for Predator or other reconnaissance aircraft to set up surveillance on the house and watch for activity. If we spotted anything suspicious—say, four or five insurgents coming and going with AK-47s or rocket-propelled grenades (RPGs), which certainly set off a red light—we marked the spot for special attention by Uncle Sam's ordnance depot. Either that or we used it as a pivot point to work on discovering the rest of the network. That generally entailed following people in vehicles around the city. Predator could tail them wherever they went—and they never seemed to catch on.

One morning when I showed up for my shift at the GCS, cup of coffee in hand and fresh from out of a warm bed with my wife, Capt. "Snowy" Middleton, the off-going pilot, was tracking an older-model Suburban through the pre-dusk streets of northwest Fallujah in the Jolan district.

"I've been following the goofy bastard all day," Snowy said. "We think he's part of an IED [improvised explosive device] team—but so far we haven't seen him do squat. Blast his ass if you get a reason."

I had yet to blast anyone—and it wasn't in the cards for me to blast this guy either. At least not today. He went home and stayed all night; the time difference between Iraq and Nevada was about twelve hours, which meant that when the sun rose over Nevada, it was sinking in Iraq. He was still holed up, apparently snoozing away, when I went off-duty and headed home to the little house I shared in Las Vegas with Trish.

Like a cop on his beat, I grew increasingly suspicious on-duty of everyone and everything. I wasted an hour in orbit one day studying a satellite TV dish on a rooftop trying to decide if it might be a camouflaged mortar tube or a rocket launcher. I trailed an old rusty Ford through heavy morning traffic as the driver went about his errands, some innocent, others possibly nefarious. I spied a group of young men hanging around on street corners shooting the breeze like American kids in Chicago or Los Angeles. I watched some men in a courtyard around a house unloading a bongo truck. Even with the longest lens in the world and excellent computer screen resolution, it was often hard at ten thousand feet to distinguish contraband from legitimate cargo.

Anytime insurgents fired a rocket or a mortar round from a location, intel marked the particular site for future reference. We checked it again and again throughout a shift to see if the enemy might return to the scene of the crime. I combed a park one night for three hours on that expectation, but, as so often happened, nothing happened.

Insurgents were always sneaking about alleys and rooftops to take potshots at U.S. troops manning roadblocks and checkpoints that closed off thoroughfares to and from the city. One night, marines issued what police would have called a BOLO (be on lookout) for a faded green Toyota pickup with a .50-caliber machine gun mounted in its bed. That was certainly a distinguishing characteristic. It had sped past a marine checkpoint, pounding away with the gun. Marines returned fire, but the Toyota and its ragtag load of fighters escaped.

For the next three days, crews at Nellis searched relentlessly for some sign of the Toyota. One afternoon, something out of place in a vacant lot in the southern industrial district of the city caught my eye. I circled. The lot contained a few pieces of broken farm equipment and an abandoned ambulance. What attracted my attention, however, was a small pickup truck partly covered with a tarp. The Iraqis covered only things they didn't want us to find or recognize.

Imagery experts who monitored our video feed compared the photo of the tarp-covered truck with the satellite photo snapped of the Toyota in action a few days earlier. Enough of the pickup was exposed to make a positive identification. Back came the response through my chatroom.

Confirmed. It's a match. That's our guy.

It was up to ground commanders in-country to make decisions on what action to take. I was in support of the intelligence analyst imbedded with Fallujah marines. He obtained permission from the J-3 chief of operations at the Combined Operations Center in Baghdad to eliminate the target. My excitement began to mount. This would be my first opportunity to employ live ordnance against the enemy. Lasing targets for fast movers or an AC-130 gunship didn't count. Blowing up the Toyota would send a message to the insurgents that they needed to do a better job hiding their stuff.

The MQ-1 Predator at that time could deploy three different types of missiles. The M-Model Hellfire with the nine-millisecond delay fuse was designed to bust through concrete walls or armor and destroy whatever was inside, incidentally limiting damage to surrounding structures and people.

The K-Model Hellfire was originally designed for helicopters as another way to attack tanks and other armored vehicles. Its two shaped charges directed its explosive energy into a single point moving forward with the missile. It first punched through the toughest armor, then exploded inside the vehicle. A nasty piece of ordnance.

We called the third Hellfire "Special K," a regular K Model with an even nastier antipersonnel bonus. When the two charges, wrapped in a sleeve of scored steel, detonated, the sleeve shattered along its score lines and blasted out razor-sharp shrapnel in all directions to slice and dice anyone within a twenty-foot radius (depending on the surface). Even those out to fifty feet might not escape its wrath.

A year later we would field a new version of the missile—the Hellfire P. This missile was taught to talk to the Predator so that it knew where the laser was pointing—giving it the ability to come off the rail, then turn and acquire the laser spot. This allows much more flexibility in maneuvering the aircraft.

But at the time I carried an M under one wing and a Special K under the other. I selected the M to hurl against the unoccupied truck and flew the Predator out to a distance of about ten klicks (kilometers) before turning inbound to start my target run. The optimum altitude to launch a missile was ten thousand feet, the range distance window for release between seven and a half and nine klicks. Too near the target and the missile would likely overfly it. Too far away and it would run out of energy and hit short.

I powered up the missile and tested it while my sensor operator in the next seat locked on to the target. My palms were sweaty and my heart pounded. Flight Lieutenant Gambold, one of our attached British pilots, hurried into the GCS trailer to act as safety observer. He stood behind my seat, bent over my shoulder with his gaze glued on the monitoring screen. I barely noticed, I was so intensely focused on the Toyota in my crosshairs, waiting another second or so to close in before release. I wanted the shot, my first, to be a good one.

The laser was locked. I took a deep breath. Once a missile was ignited, it couldn't be turned off or recalled.

At the last moment, a man unexpectedly appeared on the HUD, walking across the vacant lot and up to the truck, glancing about surreptitiously as though he felt someone watching but couldn't see anyone. I wondered if a mouse might not feel like that just before a hawk dropped out of the sky to snatch it up with piercing talons.

Poor bastard. It looked like he was about to get into the Toyota and drive off to pick up his buddies and go shoot up some more marines. Or that might at least have been what he was thinking.

Call him a bonus. Truck and driver. Blue light special, Kmart shoppers. Two for the price of one.

My right index finger tightened on the control stick trigger, the "pickle." I was concentrating entirely on the shot and its technical aspects. Right range, right speed, locked in. The man wasn't *really* a human being. He was so far away and only a high-tech image on a computer screen. The moral aspects of it—that I was about to assassinate a fellow human being from ambush—didn't factor in. Not at the moment. Not yet.

Suddenly, just before I fired, the screen froze. Alarms began sounding and warnings flashed on the heads-down display. There had been a power outage in the satellite uplink in Europe. I had no further control of the plane. It automatically shut down its lasers and weapons and turned away from the target to fly a lost link profile.

It took several minutes to get things sorted out and regain control. By then, the aircraft was flying its programmed pattern several miles away. By the time I flew back to the target, the truck was gone and I couldn't find it again. Although I may have been somewhat relieved that I hadn't had to kill a man today, I was far more frustrated that an enemy had escaped and was, at that moment, possibly on his way to kill Americans. A technical glitch had robbed me of my first shot. That was one lucky bastard down there.

Several days later, the Toyota reappeared on the vacant lot, covered up with the tarp as before. One of the on-duty Predator crews took it out. The abandoned ambulance next to it got beat up a little. By the time the locals spun it to the Arab media, they had the ambulance fleeing with a load of women and children when Americans bombed it.

Nellis Air Force Base covered nearly twelve thousand acres of scrub and desert in southern Nevada, the main post only seven miles northeast of the central business district of Las Vegas. Trish and I had our own house on the northwest side of town. I lived a schizophrenic existence between two worlds, one as a combat pilot fighting a war halfway around the world, the other as an ordinary American citizen. No one who saw me off-base and out of uniform would have ever guessed what I did for a living.

Each working day, like a commuting factory worker or a retail salesman, I reported to the POC for my shift where a sign greeted me with *You Are Now Entering CENTCOM AOR*. It could just as easily have read *You Are Now Entering C. S. Lewis' Narnia* for all that my two worlds intersected.

My first shift out of training was midnights, which meant daylight in Iraq. I drove in at 2300 hours (11:00 p.m.) and received my briefing at the POC briefing room before I strode casually into the GCS, pausing briefly along the way to exchange a few words of greetings with friends and fellow workers and pick up a cup of coffee. Then I took my seat at the Predator's controls and became immersed in warfare for the next six to eight hours.

The sun was coming up by the time I drove home again, perhaps stopping along the way to pick up a newspaper or something from the store for Trish. It could be disorienting, a real disconnect, to live simultaneously in two such different worlds.

The evening after I almost shot the Toyota pickup was my night off. It had been a nerve-wracking shift. I grabbed a few hours' sleep before Trish and I drove to the strip for a good steak. Seen from space, Sin City, the "city that never sleeps," where "what happens here stays here," was the brightest spot on earth. I hardly noticed all the lights coming on against the approach of nightfall. Trish observed that I was abnormally quiet, introspective, playing with my food or staring off in deep thought.

"Out with it, Matt," she finally prompted. Sometimes it was like she read my thoughts. "What happened?"

I took a deep breath. "Trish . . ."

Her nickname, Ruby, seemed appropriate, what with her red hair and blue eyes, but I normally called her "Trish," leaving "Ruby" for the rest of the family to use.

"Trish, I almost killed a man today." I hesitated before telling her the rest. "Trish, I'm sorry I didn't kill him."

Anyone else would have been shocked at such a confession. Trish merely waited without comment while I briefly told her about how I almost took out the Toyota and its driver, prevented from doing so only because I lost the satellite link. Then I fell silent. She didn't press. She was not only my wife, she was also my best friend with whom I could discuss anything. She knew when to push and when to let me sort out things in my own way. I would talk it all out when I was ready.

I wasn't ready. Not yet. Things were still a bit confused in my mind. Dad always said I had a way of analyzing the life out of things. He liked to tease me by calling me "the Professor," but there was a ring of truth to it. I was a reader, especially of history and philosophy. From such readings and from my own limited experiences so far, I could not but realize that mankind was entering a new era of warfare for which neither history nor philosophy completely prepared us. Profound questions confronted not only me personally but our entire society. What laws and ethical codes applied, I wondered, when men sent out unmanned machines to fight for us?

War, the idea of warfare, was as old as man himself. I abhorred it. Yet what most concerned me personally was that I had considered none of this as I rushed to pull the trigger and annihilate another human being. After all, mission came first. I was beginning to discover that war touched a place in us, as John Keegan noted in *A History of Warfare*, "where self dissolves rational purpose, where pride reigns, where emotion is paramount, where instinct is king." I sometimes pondered how Adam might have gone back to the Garden of Eden and whacked the serpent. That was what I would have done.

I was almost ashamed to admit to Trish the thrill I felt at the moment I prepared to squeeze the trigger. My thoughts had been fundamentally superficial. I had trained through dozens of practice target runs. I knew how things were supposed to go—and I was ready. I operated solely through my training, focused on one thing. Instinct was, indeed, king.

It had not been quite real, even afterward. I was among the first generation of soldiers working with robots to wage war. The ability to kill people from such great distances, playing God, widened the gap between the reality of war and our perception of it. It was almost like watching an NFL game on TV with its tiny figures on the screen compared to being down there on the field in the mud and the blood in the rain getting your socks knocked off. "The pleasures of a spectacle with the added thrill that it is real for someone, but not the spectator" was how analyst Christopher Coker expressed it. All the potential gains of war without the costs. It could even be mildly entertaining. Could it not also become too easy, too tempting, too much like simulated combat, like the computer game Civilization?

I experienced the psychological disconnect of being "at war" for eight hours or so, shooting weapons at enemy targets, directing hits against the other side, and then I got in my car, drove home, and, that evening, had steak with my wife in Las Vegas. I had wanted to do more while flying RC-135s out of Crete at the beginning of Iraqi Freedom. I was now doing more. Within the next year, *more* would entail deploying to Iraq to work my rotation at the Launch and Recovery Element (LRE). That meant getting down on the field closer to the mud and the blood and the rain. Taking my own risks and confronting my own demons, as all soldiers must.

Trish had a way of getting directly to the heart of a matter, of encapsulating its essence in the simplest possible manner. She reached across the table and took my hand.

"Matt," she said, "what you're doing is saving lives. Not just American lives but Iraqi lives as well. If it weren't for you and the other pilots, troops on the ground would have to do the job—with far greater damage and far more carnage."

I squeezed her hand. "This is why I love you," I said.

"Eat your steak, Professor."

CHAPTER 5

Rocket Man

NAJAF WAS THE HOLIEST city of the Shiite Muslims because of the Imam Ali Mosque—the Mosque of the 13th Imam. Baby-cheeked leader and cleric Muqtada al-Sadr had chosen to make his stand at and around the mosque in the ghetto called "Sadr City," named after Muqtada's grandfather, also a cleric. Our generals had declared the mosque a no-fire zone and therefore, as an unintended consequence, a haven for every would-be martyr in Najaf or Baghdad. Al-Sadr's Mahdi Army—*mahdi* being the Arabic term for messiah—believed the cleric to be the messiah returned to earth from the supernatural world to destroy the infidels.

While al-Sadr and most of his militia holed up in the mosque, a few of his fighters were unwilling to wait for the infidels to come to them. The "Rocket Man," as we dubbed him, was the most notorious of these lone wolves. Like a rat, he slithered through the slums of Sadr City armed with 100mm supersonic rockets equipped with 5-pound high-explosive warheads, killing and maiming GIs, marines, and Iraqi bystanders. He wasn't *that* accurate with his rockets, or apparently too particular who he targeted, which accounted for his high rate of collateral damage against civilians. He must have trained with *The Gang That Couldn't Shoot Straight*.

The Rocket Man quickly rose to the top of our Most Wanted list. We finally got lucky. Either that or he got careless. He popped a rocket at a U.S. Army squad patrolling an alley. He missed and took out the front of a nearby house. What he didn't know was that I was watching through the camera of the Predator soaring ten thousand feet above his head.

"We finally got the perp!" I exclaimed.

The Rocket Man jumped up and down in frustration at having missed. Then, before the army squad could zero in on him, he tossed his launcher into the trunk of a Ford compact rattletrap and hauled coal, driving like a maniac down alleys and side streets, taking corners on two wheels and generally scaring the hell out of pedestrians and dogs who got in his way. I followed. He had created a lot of chaos and destruction during his reign. I was determined to bring it to an end.

Striking a moving target with a Hellfire posed an almost insurmountable challenge because of the two-second control delay inherent in the satellite link. My best bet was to wait until the Ford stopped, hopefully in an open area where my sensor operator and I could whack the guy without injuring bystanders. But whatever it took, I acknowledged grimly, we *were* going to kill him.

Orbiting high over Sadr City, I kept my camera trained on the old Ford as it weaved and bobbed toward some unknown destination. The live-action computer screen in front of me provided a movie-like image of sun glaring down on sand-colored buildings, narrow streets teeming with life, smoke rising north from where marines were engaged. Rocket Man knew the city well. He avoided traffic congestion, roadblock checkpoints, and main streets where he might encounter U.S. Army or Iraqi Army (IA) patrols.

It didn't matter how cautious he was. I had my eye on him. His hours were numbered. Even if he thought to look up, chances were he would never spot the speck that the Predator made high in the sky against the bright desert sun.

I was on the radio and in the chatrooms with the on-ground forward air controller, also known as a joint terminal attack controller (JTAC), feeding him live video so he could see what I was seeing. He was ready to grant clearance as soon as I had the opportunity to engage.

To my disappointment, the Ford maneuvered into one of the densest sections of Sadr City. The driver slowed to avoid hitting people, sometimes stopping altogether to allow burqa-clad females to cross in front of him. Judging from his past behavior, I doubted he was particularly solicitous of human life. He just didn't want to draw attention to himself.

The neighborhood was jammed hip to shoulder with little flat-roofed houses the color of the desert, each with its own tiny, bare courtyard. Kids, stray dogs, and rats rummaged through piles of garbage and among junk cars jacked up on blocks in the streets.

"Keep going," I urged the Ford under my breath.

But of course it didn't. It slowed in front of a little house enclosed by a high mud-brick wall and sat there with engine idling for a few moments while the driver checked to see if he was being followed. Then, apparently satisfied, he eased through an opening into the courtyard and parked in the shade beneath a giant eucalyptus tree, leaving only a patch of rusty tan roof visible from the air. I couldn't make out whether or not he got out of the car and went into the house. I didn't think he did.

"Houston, we have a problem," I transmitted to JTAC.

My radio crackled. *"Roger that, Agony 31."*

Glued to the video feed I supplied him, JTAC saw and recognized the situation as well as I. We had to be cautious with a shot in this neighborhood to avoid killing a bunch of people who didn't necessarily deserve being killed. At the same time, we weren't about to let this savage get away from us.

I brought the airplane around to get a peek underneath the tree. Best I could tell, the guy was still sitting at the wheel waiting for the heat to lift off so he could go back out and try another shot at a different location. He might not miss the next time.

Kids in the street chattered past the mud-brick wall, laughing as they took turns rolling an old car tire. A woman on the flat roof of the house in whose courtyard my guy hid was hanging out laundry. She cast a single nonchalant glance at the vehicle parked underneath her tree. The Rocket Man probably didn't live here, but the woman more than likely knew who he was and why he was laying low.

Judging from the heat signature on my infrared screen, a number of people occupied the house next door. Other people were either walking or were stationary in groups all up and down the street, like a summer day in Garden City, Kansas, or any other small town in the United States where the living was easy.

Leave it to me to spoil their day.

I carried two Hellfires—one the Special K Model that detonated upon impact and directed its force outward at the point of impact, the other an M with a time delay fuse that punched through a target before blowing up and out. Nothing could ruin a tank driver's day like an M penetrating his armor and detonating inside his tank. But since the M was likely to go right through a soft-skinned vehicle like the Ford and expend its energy deep in the soil, I selected the K Model as my best chance to destroy the vehicle, driver, and rocket tubes.

I was nevertheless hesitant about firing. The thought of living in the aftermath of having harmed or killed innocent people chilled the marrow of my being. The JTAC and I had to decide together whether the payoff was worth the risk. We were comforted in the knowledge that the brick wall surrounding the vehicle should contain much of the blast energy.

The JTAC finally gave me the "cleared hot" call. However, he advised me to wait as long as possible on the off chance the guy would move to a more suitable location.

I continued to orbit, waiting, my camera focused. Uncertainty crawled in my stomach like a bed of worms. I had never killed before, but the Rocket Man *deserved* being killed if anyone did.

Nonetheless, I couldn't help feeling like a villain sneaking around waiting for the right moment to commit perfidy.

After an hour or so, it became clear that the guy was going to stay in his hole until at least nightfall, by which time I would be running out of fuel and have to pull off. It was now or never.

I began preparations for a shot by scrutinizing the target from all angles in order to choose the best approach to minimize collateral damage. I calculated that if I dropped one right down the middle of the yard on top of the Ford, the brick wall would buffer the explosion and leave adjacent houses relatively undamaged. Nobody else should be hurt, which was an integral element of our rules of engagement. I doubted whether B-17 and B-29 pilots and bombardiers of World War II agonized over dropping tons of bombs over Dresden or Berlin as much as I did over taking out one measly perp in a car.

I flew the Predator out to about twelve klicks, then turned inbound for the run, powering up the K Model at the same time. Senior Airman Juan Abado, my sensor, armed his targeting laser. It was his job to guide the missile to its target once I fired it. If his hand twitched at the last instant, if he breathed wrong, the missile might go astray and take out the house full of people next door or the group of old men smoking and joking down the block.

We were on final at ten thousand feet and eighty knots, flying within two degrees of missile-to-target heading. At 10 klicks out, I ordered Abado to begin lasing. At 8.9 klicks, I initiated my three-second countdown to "rifle" at 8.7. The moment of truth was upon me. My first kill.

Concentrating on crosshairs superimposed on the target, I drew in a deep breath, felt sweat stinging my eyes, tasted the bile of excitement in my mouth. I took a last look at the street in front of the house to make sure it was clear. Then I squeezed the trigger. The image on my screen pixilated as the airplane yawed from the asymmetric thrust of the missile's launching.

It would take about thirty seconds for the missile to reach its target. I began a silent countdown, watching my screen, waiting for *splash!*

At the last moment, with only seconds to spare, the unthinkable occurred. An elderly man appeared on the screen, tottering along in front of the wall in his traditional Arab garb. He was already about two-thirds of the way past, which put him directly outside the wall from the Rocket Man's parked car. There was no way to recall or divert the Hellfire without creating even more devastating unintended consequences.

"Oh, Jesus! Move!"

My infrared screen flashed bright as the heat wave from the explosion washed across camera lenses. The image re-formed almost instantly to reveal that the missile had detonated precisely where I aimed it. Flames from the car were already crackling into the tree branches under which it was parked.

And then I saw the breached wall with bricks blasted into the street. I looked for the old man, who must have been the most unfortunate SOB in Sadr City because he just happened to be walking by at the wrong time. I saw that the shock wave had carried him into the middle of the street, where mobs of people swarming in the aftermath hid him from my further view. I was unable to determine whether or not he got up.

As for the Rocket Man, I doubted if even seventy-two virgins in Paradise would ever be able to put him back together again. Low on fuel, I turned the airplane toward home and then passed it off to the LRE at Balad for landing and refueling. I sat silently, contemplating my console and screens for a few minutes, as did Airman Abado. The Rocket Man had it coming. The old man did not.

By the time this war was over, I later reflected during dark moments, I was apt to have more innocent blood on my hands. Innocent blood on my hands, rubble and wreckage in my wake, and Iraqi mothers and wives cursing me—or the idea of me—and praying for my damnation.

Those who would call this a Nintendo game had never sat in my seat. Those were real people down there. Real people with real lives.

I still wasn't ready to discuss it with Trish.

In Iraq, Predator flew mostly over cities, such as Baghdad or Fallujah. Afghanistan, on the other hand, was open and rural; it took a long time to get anywhere at a hundred miles per hour. The contrast of flying Iraq one day and the vast wastelands of Afghanistan the next was almost a cultural shock. The Operation Enduring Freedom job was duller than watching turtles race at a county fair back in Indiana. Most of the time I stared down at rocks, dirt roads, trails, mountains, and the occasional farmhouse or remote village.

Things began to look up during one shift when we received intelligence that might lead us to The Man himself—Osama bin Laden. The source pointed to a tiny village along the Afghanistan-Pakistan border. Bin Laden, it seemed, preferred to dwell in places where few other people would go. I headed that way, throttle open, along with a couple of A-10s, a B-1B bomber, and a dozen marine helicopters. Excitement at the prospect of capturing the most wanted terrorist in the world crackled over the secure radio net.

I was actually in pursuit of Osama bin Laden! How much closer than that could one get to the war?

In many ways, the wars in Iraq and Afghanistan led to the doorstep of Osama bin Laden, founder of the worldwide terrorist organization known as al Qaeda. He was responsible either directly or indirectly for the deaths of thousands of people, Muslims and "infidels" alike. In his name, radical Islamists had bombed embassies, cruise ships, passenger airliners, nightclubs, the USS *Cole*, and New York's Twin Towers. They also organized and fought as insurgents in Somalia, West Africa, Algeria, Palestine, Pakistan, Iraq, Afghanistan. If President George W. Bush proved correct in his assumption that the War on Terror might continue for decades, bin Laden could well take his place among the likes of Hitler, Mao, Stalin, and Pol Pot as one of the world's most virulent figures.

The son of a wealthy Sunni businessman and his father's tenth wife, bin Laden was born in Saudi Arabia on March 10, 1957, and grew up a devout Wahhabi Muslim with close ties to the Saudi royal family. He studied economics and business administration at the secular "elite" Al-Thager Model School in Riyadh, where he also wrote poetry and dabbled in music. A tall, gaunt figure at six feet five, he eventually married four women and fathered at least two dozen children. Muslim men were allowed multiple wives, their numbers signifying a man's wealth and prestige.

With that many wives and children, I wondered how he had time to become a terrorist.

Radicalized through his strict Wahhabi upbringing, he came to believe that only the restoration of Sharia law and its expansion throughout the world would set things right. He settled on expansive jihad as the single most effective method to right perceived injustices against Muslims perpetrated by the "Great Satan" and by Israel, the "Little Satan." America was the Great Satan due to, in his words, "intoxicants, gambling and usury and . . . immoral acts of fornication and homosexuality." The Jews were "masters of usury and leaders of treachery. They will leave you nothing, either in this world or the next." The most outrageous part of his ideology was that civilians—women and children—were legitimate targets of jihad, thus the bombing of schools, school buses, and pizza parlors. His perception of Islam was so stern that he even opposed instrumental music and kite flying.

He left college in 1984 to fight against the Soviet invasion of Afghanistan. Afghanistan under the rule of Mullah Omar's Taliban was, he believed, the "only true Islamic country" in the Muslim world. It was then he had his "coming out," emerging as the ultimate jihadist in his military camouflage jacket with a captured AK-47 assault rifle slung across one shoulder. Urban legend had it that he obtained the rifle by killing a Russian soldier with his bare hands. He was variously referred to throughout the Muslim world as Prince Osama, the Lion Sheikh, the Director, or the Samaritan. Even the Saudi royal family hesitated to denounce him.

Bin Laden returned to Saudi Arabia in 1990 as "the hero of Jihad who had brought down the mighty [Soviet Union] super-power." At the height of his career, he refused to rest on past laurels. He promptly declared a worldwide jihad against the West and established terrorist training bases in Albania, the Sudan, and Afghanistan. He assisted jihads financially and sometimes militarily in Algeria, Egypt, Ethiopia, and Palestine. The first bombing attack in which he was believed to be involved occurred on December 29, 1992, when the Gold Mihor Hotel in Aden was blown up, killing two people.

In 1997, he funded the November massacre of sixty-two tourists in Luxor, Egypt. In March 1998, Libya issued the first international arrest warrant for him and three other terrorists for murdering two German citizens in Libya. In June 1998, the United States indicted him for the 1995 truck bombing of the U.S.–operated Saudi National Guard Training Center in Riyadh. Americans indicted him again in 1998 for master-minding the bombings of U.S. embassies in Kenya and Tanzania. That propelled him to the FBI's Most Wanted list and made him the most hunted fugitive on the globe.

Bin Laden and Ayman al-Zawahiri, his second in command, cosigned a fatwa against Jews and other "crusaders" in the name of the World Islamic Front for jihad. In it, they declared the killing of North Americans and associated allies an "individual duty for every Muslim." Americans, bin Laden said, were "very easy targets . . . You will see the results of this in a very short time."

President Bill Clinton authorized the CIA to apprehend bin Laden and bring him to justice, dead or alive. In August 1998, sixty-six Cruise missiles launched by U.S. Navy warships in the Arabian Sea descended on bin Laden's terrorist training camp in Afghanistan, missing him by only a few hours. A CIA scheme to infiltrate his Pakistani command post to capture or kill him in 1999 failed because of a coup against the Pakistani government. In 2000, CIA-funded foreign operators fired RPGs at a convoy of vehicles in which bin Laden was traveling through the mountains of Afghanistan, hitting one of the vehicles but not bin Laden's.

He seemed to be living a charmed life.

After the 9-11 attacks against New York City's World Trade Center and the Pentagon, bin Laden released a statement saying, "God knows it did not cross our minds to attack the towers, but after the situation became unbearable and we witnessed the injustice and tyranny of the American-Israeli alliance against our people in Palestine and Lebanon, I thought about it. And the events that affected me directly were that of 1982 and the events that followed—when America allowed the Israelis to invade Lebanon. As I watched the destroyed towers in Lebanon, it occurred to me to punish the unjust the same way and destroy towers in America so it could taste some of what we are tasting and to stop killing our women and children."

Nineteen hijackers, most of whom were from bin Laden's home country of Saudi Arabia, seized four American airliners—United Airlines Flights 93 and 175, and American Airlines Flights 11 and 77. They crashed two of them into the Twin Towers, and another into the Pentagon. Passengers on the fourth airliner resisted; the plane prematurely crashed into a field in Pennsylvania, killing all aboard. Including the 19 fanatical martyrs, this "act of war" killed 2,974 people. Bin Laden made the top of the list on the FBI's Most Wanted Terrorists.

Less than a month later, the United States declared its "War on Terror" and invaded Afghanistan to depose the Taliban regime and capture bin Laden and his al Qaeda cohorts. A year and a half later, the United States invaded Iraq on the grounds of "weapons of mass destruction" and Saddam Hussein's alleged support of al Qaeda terrorism.

Now, in 2004, flying the Predator at maximum cruise speed across Afghanistan, I felt as if I were on a rendezvous with destiny. To be in on the destruction or death of one of the most evil forces on the planet was a tremendous responsibility and a great honor. Evil rarely died a natural death. Osama bin Laden had escaped justice time after time. Maybe his luck had finally run out.

CHAPTER 6

Fighting for Millennia

BECAUSE I WAS UNARMED in Afghanistan, taking out bin Laden myself was not an option. The best I could do was to keep an eye on the village in the event that our pigeon was hiding there with some of his wives and concubines—and possibly coordinate the attack if he were found there. I supposed he took with him into the war zone only those who could be considered disposable. If he tried to get away, I would track him into the mountains with my spy-in-the-sky eye, whether he fled by auto, mule, camel, or shank's mare, and direct fighter-bombers down on his murderous head.

"*Pacman One-Two, our ETA is estimated at two-zero mikes. Are you on-station?*" That was a marine JTAC in one of the Blackhawk choppers filled with assault troops.

"A-firm, Leatherneck," I replied. "I'm arriving overhead now." Actually, *I* wasn't *arriving* anywhere. I was in Nevada.

"*Pacman One-Two, say when ready for check-in.*"

I delivered the standard check-in brief to the JTAC. There wasn't much to say in the way of a situation report. While excitement electrified the airwaves above the village, the settlement itself was like timeless hundreds of others in the region as they began to stir for the day. A scattering of a dozen or so mud-brown huts were tucked into one end of a valley at the base of a rock-strewn mountain tinted with the pink-orange of a sun rising through clouds. Horses and camels lolled with their heads hung over pole corrals. The archaic scene on my computer screen struck me as wry irony. The most dangerous international criminal in the world, who had terrorized a great nation and brought it to war, was living in primitive squalor not far removed from the Stone Age while being pursued by space age technology. For over two decades, this man had succeeded in eluding first the Russians and now virtually the entire free world.

"Leatherneck, this is Pacman One-Two," I radioed, studying my screen for movement and finding none. "It appears all quiet—"

My video suddenly washed out in a mass of swirling gray. I knew immediately what I had done. I had been so "locked on" the target, with my cameras focused on the village, that I inadvertently flew the Predator into a cloud bank. At lower altitudes or in the dry desert air of Iraq, it might not have been cause for panic. However, at twenty thousand feet altitude, the average temperature above the Hindu Kush Mountains was around minus twenty degrees Celsius. Add moisture to the equation—and clouds were moisture—and I had a crisis. Ice would begin forming on the wings instantly, increasing the weight of the airplane and spoiling its lift. Enough ice and the plane would go down.

Most other aircraft designed to fly at these altitudes came with deicers such as inflatable boots that popped ice from the wings or a plumbing system that leaked deicing fluid onto them. Earlier versions of the Predator had had such equipment, but it had been removed in favor of carrying more fuel in Afghanistan and missiles in Iraq.

The last time one of our pilots flew into icing, the plane vanished and was never seen again. I knew what was coming: A board of inquiry, safety investigations, endless questions that reached one conclusion: pilot error and possibly the termination of my Predator-flying days. The loss of a four million dollar machine was not chump change. I had a cruel vision of spending the rest of my career in a missile silo in the Middle of Nowhere, Wyoming.

I could beat myself up over it later. Right now, my only hope was to exit the clouds ASAP. I executed a blind 180-degree turn, hoping to get out the same way I came in. My heart was pounding.

The HUD suddenly flipped from swirling cloud gray to—*frozen*. The video stopped moving, indicating a lack of feed. I had lost the satellite link—the same thing that happened over Fallujah when I was about to squeeze the trigger on the Toyota pickup. The clouds must have interfered with the signal. My control inputs were useless. I was no longer a pilot. Not even a passenger. The airplane was now on its own, flying a "lost link" pattern that would likely drive it deeper into the clouds while more and more ice accumulated on the wings until it crashed. Middle of Nowhere, Wyoming, might be a step up in my career.

And Osama bin Laden right underneath and perhaps getting away again.

Senior Airman Mata, my sensor operator, was glued to his black screen, as fully alarmed as I, his lean face looking pale and stricken. "Jesus, Jesus . . . ," he murmured in a continuous litany.

Suddenly, to my surprise and relief, video unexpectedly returned. I had control of the aircraft again as we broke free of the cloud bank and entered a little pocket of clear air surrounded by building thunderstorms. I told Mata to rotate the MTS pod and looked at my wings, discovering about a half inch of rime-ice buildup on the left leading edge. The plane felt sluggish.

Osama bin Laden or not, the mission was over for us. We either went down in a ball of ice—or we found dry air somewhere else. Either way, I acknowledged with growing frustration, I wasn't going to be in on capturing one of the world's most wanted terrorists.

I swung the plane around, looking for a path out of the clouds.

"There!" Mata exclaimed. "Three o'clock."

Sure enough, to our three o'clock appeared a gateway between columns of boiling gray-and-white thunderheads. I banked toward it, holding my breath in anticipation that it would lead into clear air and not into a blind end. For several minutes, neither Mata nor I dared even to hope as I soared between clouds rising thousands of feet off the tip of each wing.

Blinding sunshine washed across the screen as the plane broke free. Within a few minutes, the ice melted and we were back to normal operations. Except that I had to inform the units we were supporting that we couldn't fly the north corridor today—Osama bin Laden or not. Call it fate, call it luck, at least my air force future had been saved.

I monitored the radios as the raid launched, even though I was no longer a participant. Marine helicopters swooped down from the sky, catching the village by surprise and surrounding it with troops. Camels watched with their dopey expressions as leathernecks searched house to house. The raid turned out to be anticlimactic—another dry hole.

At least we knew one more place where Osama bin Laden *wasn't*.

I always liked to keep a number of chatrooms open on my control console in order to feed my curiosity with greater insight into how the War on Terror was going. In one window, for example, JTACs discussed things they'd like to blow up and how they intended doing it. In another, generals at Central Command (CENTCOM) or the Combined Air and Space Operations Center (CAOC) discussed "big picture" issues such as strategies and the broad-range use of troops and armies. The irony of the discussions never failed to impress upon me the fact that Americans were only the latest people to have fought and died in Iraq's Tigris-Euphrates region, "the cradle of civilization."

As a bit of a bookworm, I liked to think of myself as a classic scholar-soldier. Sometimes, even Trish called me by Dad's old nickname, "the Professor." I received a proper education in regional history and geography during my shifts flying the Predator from the GCS at Nellis, soaring slowly above Iraq for long hours at a time, snooping and sniping, waiting for a mission or riding one out. Along with my off-duty reading and area studies, it afforded an unprecedented opportunity to ponder cities, sites, and shrines, some of which were already ancient in biblical times.

The written history of Iraq went back to before the Old Testament of the Bible, the oral history back even further to the days of Sargon the Empire Builder. During its last five thousand years of hosting empires, of invading and being invaded, the fertile valley between the rivers had suffered many rulers— Sumerian, Assyrian, Persian, Greek, Roman, Mongol, Turk, and British, among others, leaving behind such ancient ruins as Babylon, Ur, and Nineveh. Some of the world's greatest civilizations developed in this area. Between the seventh and thirteenth centuries, Baghdad was renowned for its scholars and artists when few people in Europe could even read, much less write.

If the Bible was to be believed, mankind began in the valley. Some say Iraq was the approximate location of the Garden of Eden where, in the beginning, God created Adam and Eve. While Israel was cited more times in the Bible than any other nation, Iraq followed in close second place under such names as Babylon, Land of Shinar, and Mesopotamia. The River Euphrates was mentioned in Genesis, the first book of the Old Testament, and in Revelation, the last book of the New Testament—and fifteen other times in between (in the King James version).

Satan made his first cited appearance in Iraq. The Tower of Babel was built in Iraq, followed by the resulting confusion in languages. Abraham hailed from a city in Iraq, as did Isaac's wife. Jacob spent twenty years between the two rivers. Iraq and Iran were the site of the Persian empire. The greatest Christian revival in history occurred in Nineveh, now the modern city of Mosul. The events in the book of Esther took place in Iraq. The book of Nahum prophesied against a city in Iraq. The Euphrates River was the far eastern border of the land that God promised to Abraham. The prophet Muhammad, founder of the Muslim religion, was born in Mecca in what is now the nation of Saudi Arabia. Finally, the Book of Revelation warned against the resurrection of Babylon from which, according to Biblical prophecy, the Antichrist would rule. Saddam Hussein had undertaken the task of restoring the ancient city before he was deposed by the U.S.–led invasion.

In 586 B.C., Nebuchadnezzar II, credited for building the legendary Hanging Gardens of Babylon, conquered Judea (Judah) and destroyed Jerusalem. Solomon's Temple was also destroyed. It is said that Nebuchadnezzar II carried away over fifteen thousand captives and sent the rest of the population into exile in Babylonia. Iraq had been conquered more times during its history than practically any other country in the world. The Persians swept it into their empire in 550 B.C., losing it first to Cyrus the Great in 539 B.C. and then to Alexander the Great in 331 B.C. The Greeks called it Mesopotamia, "Land Between the Two Rivers."

It became part of the Roman Empire in A.D. 115 and remained under Roman rule until Arab armies conquered it in A.D. 637, bringing the Arabic language and the Muslim religion. For the next six hundred years, Baghdad, as capital of the Arab empire, reigned as a center of learning known for its achievements in the arts and sciences. But after that it was downhill.

In 1258, Mongols invaded from the east and cast the region back into the Dark Ages, a collapse from which it never fully recovered. History had asked tribal Arabia to adjust to modernity in less than a century, a task that took the West nearly six centuries to accomplish.

Iraq suffered another series of upheavals after the Ottoman Turks ran off the Mongols in 1534. The British seized Mesopotamia from the Turks during World War I and governed it under a League of Nations mandate until it was granted independence in 1932. When World War II erupted, Iraqi Army officers wanted Iraq to join Hitler and the Axis powers. However, Britain defeated the Iraqi Army and drove pro-Axis leaders from the country. Under new rule, Iraq dutifully declared war against Hitler in 1943. In 1945, it helped form the Arab League, which promptly set itself against the creation of the State of Israel.

Attacks by the league against Israel resulted in defeats that led to uprisings in the Iraqi government in 1948 and again in 1952. In 1958, army officers overthrew the monarchy under King Faisal II and replaced it with a "republic" that was in fact a military dictatorship. General Abdul Karim Qasim, who led the revolution, became premier, reversed the country's pro-West stand, and began siding with the communist countries.

Coup followed coup until 1968 when an army officer named Ahmed Hasan al-Bakr reestablished the Baath government that had ruled briefly in 1963. The Arab Socialist Baath Party—Baath or Ba'ath meaning "resurrection" or "renaissance"—was founded in Syria in 1940 with the goal of "resurrecting" the glory of the Arab nation that had been destroyed by Ottoman and Western imperialism. It would do this through unifying all Arab countries against the West.

Al-Bakr resigned in 1979 in favor of Saddam Hussein, a lawyer and Baathist leader. Things after that *really* started going downhill.

Flying the Predator at twelve thousand feet above Saddam's old fiefdom, I could see a long muddy stretch of the Euphrates River cutting through the troubled land of Babylonia. The river was 2,235 miles long, originating in the mountains of Turkey to flow southeast through the great plains of Syria and Iraq before merging with the Tigris about 120 miles from the Persian Gulf. According to Revelation, it would dry up after a full-scale invasion launched by powers coming from the East. Civilization, according to the Bible, began in Iraq and would likely end in Iraq. Blood would rise to the level of a horse's bridle during the Battle of Armageddon in the End Times.

The United States had invaded from the southeast, from Kuwait. The Euphrates didn't look like it was about to dry up. But if the Battle of Armageddon began, I reflected wryly, I was going to have a bird's-eye view of it.

For the past two weeks we had concentrated our attention on the city of An Najaf, located about ninety miles south of Baghdad and the Sadr City slums—the eastern portion of Baghdad where Shiite cleric Muqtada al-Sadr had risen to prominence. Al-Sadr and his Mahdi Army remained holed up in the Imam Ali Mosque in An Najaf, where about 2,000 U.S. Marines and 1,800 Iraqi security forces surrounded them. We had orders not to attack the shrine, no matter the provocation, even though the Mahdis had chosen to use it as a firebase. Nothing could be shot into the zone, even in self-defense. The holy mosque must not be damaged. Coalition forces were held at bay by sensitivity to the worldwide feelings of Muslims.

What resulted was a tedious standoff during which al-Sadr and his militia launched indiscriminate mortar and rocket attacks in all directions, and occasional "Rocket Man" types sneaked out to wreak havoc wherever they could. Even Najaf residents were getting fed up with the constant militia bombardments and were urging marines to go in and bring it to an end. How long that good will would last once the Mahdis were gone was anyone's guess.

I was just as frustrated as everyone else, even though I understood the Muslim veneration for the ancient shrine. It was an incredible sight to behold, even from ten thousand feet above it. The twin spires and gilded dome of the magnificent edifice glistened in the sun like burnished brass. Imam Ali, the fourth caliph, was buried there; he was the prophet Muhammad's cousin *and* his son-in-law. According to Shiite belief, the remains of Adam and Noah rested inside the mosque next to the dead caliph.

An Iranian ruler had built the mosque in A.D. 955 in tribute to the caliph. Destroyed by fire, it was rebuilt in 1086 and then again around the time Columbus discovered the New World. Following the Gulf War in 1991, Saddam Hussein's Republican Guard, showing none of the sensitivity now on display, stormed the mosque, where members of the Shiite opposition had taken refuge, and massacred all occupants. The shrine was closed after that for almost two years for repairs.

It was the third holiest site in the world for the Shiite branch of Islam. Only Mecca and Medina received more Muslim pilgrims. Over the course of more than a millennium, numerous hospices, schools, libraries, and Safi convents had sprung up around the mosque, creating an enclave of Shiite learning and theology. Not far from the dome was the Wadi of Peace, the largest cemetery in the Muslim world where several prophets were buried. The devout from other lands aspired to be buried there so they could be raised from the dead with Imam Ali on Judgment Day.

Apparently, reverence for the holy site did not prevent al-Sadr's fighters from using it to further their own ends of insurgency. On my HUD, I watched them strolling casually about the grounds, knowing as they did that they were protected. They carried weapons of various sorts—machine guns, RPGs, knee mortars. Sometimes, a man climbed to the top of a wall or onto the hood of a car and, brandishing his weapon, jeered U.S. forces outside.

Al Sadr's fighters didn't even bother to conceal a pair of mortar tubes standing tri-legged in a parking lot. At any time of the day, it seemed, right after or before the Call to Prayer, gunners strode out onto the parking lot with a sack full of ammo to pop a few rounds at marines. Afterward, they sprawled on the parking lot to have tea and smoke and joke, secure in the knowledge that no one dared retaliate.

I argued with Major "Rush" Ridenbaugh, my mission commander at Nellis, that I could easily take them out without damage to the mosque—the Hellfire missile was that accurate. But it was not his decision to make, and no one else up the chain of command was willing to take a chance.

"Captain Martin, we've been ordered to wait it out," he said, then turned and walked back to the ops center.

Defeated, I watched the goings-on down there around the mosque and wondered just how you could win a war when you weren't allowed to fight it. This mosque, a thousand years old or not, wouldn't have lasted five minutes against the U.S. 6th Marines during World War II had the Japanese taken refuge in it. I could almost hear what Gen. George Patton might have said had he and his European Third Army been reincarnated.

"Screw them ragheaded sonsofbitches. Blow that mosque back to the tenth century, then go in there and piss on the corpses."

At least, that's what I imagine Patton might say.

CHAPTER 7

Big Man in Fallujah

NAJAF AND THE CLERIC al-Sadr were only two of the challenges we had to deal with as the insurgency gained a claw hold on Iraq. The U.S. Marines remained on the outskirts of Fallujah, still not entering the city because chances seemed to grow slimmer and slimmer that it would ever return to the civilized world. The government had completely broken down. Through my Predator cameras I observed gangs of armed men prowling the neighborhoods asserting their authority through brute intimidation, erecting roadblocks to further demoralize locals and prevent outside aid from reaching them. Iraqi police had either given up on restoring order or, worse yet, had gone over to the gangs. Insurgency safe houses that we monitored from the air sometimes entertained curious visitors in blue-and-white police cars; I watched Iraqi cops carrying contraband in and out of the houses.

The big man in Fallujah, the Jordanian and al Qaeda ally named Abu Musab al-Zarqawi, had attracted a band of foreign militants in much the same way that Osama bin Laden drew outsiders to Afghanistan to fight the Soviets. Zarqawi was judged to be completely heartless, a raw psychopathic killer born for a war without moral limits on brutality. His organization, Al-Tawid Wal Jihad, "One God and Holy War," was responsible for at least seven hundred killings in Iraq since the war began. Recently, the gang began resorting to kidnapping foreigners, mostly poorly guarded truck drivers working for companies providing services to the U.S.–led coalition. At least seventy had been seized by Zarqawi and disappeared, their mangled corpses sometimes found floating in the river or half-buried in the desert. Several had been publicly beheaded on website video and distributed around the world.

In May 2004, three months ago, chilling video on a website showed five men cloaked from head to toe in black as they beheaded an abducted American civilian named Nicholas Berg. Although the executioners' faces were covered, American intelligence identified the killer leader as Zarqawi. He yanked the screaming American's head back with one hand to expose the throat and then began sawing through flesh and bone with a long razor-sharp knife in the other hand. The production ended with Zarqawi hoisting the severed head like a trophy. The video was entitled *Abu Musab al-Zarqawi Slaughters an American*.

The following month, Zarqawi similarly beheaded a South Korean civilian named Kim Sunil, followed in August by a Turkish truck driver kidnapped while transporting supplies to an American military base. Four Lebanese truck drivers vanished after that, along with a senior Egyptian diplomat. Communications received through terrorist websites threatened that all of them would be beheaded unless Zarqawi's demands were met.

I found it easier and easier to justify bombing barbarians like these back to the hell that had spawned them. I moved Zarqawi to the top of my hit list as a personal nemesis whose life I could take without a moment's hesitation and without losing a wink of sleep. In the search to run him out of his hole, I employed every spare moment not assigned to other missions. I hovered above safe houses in Fallujah for hours, watching, photographing, always on the lookout for the terrorist's stocky figure. I would often pull up his image on my mission computer to help me spot him. So many of the insurgents dressed alike, however—with their faces covered by *heffiyeh*s or baklavas, their forms indistinct in baggy clothing or robelike *thawbs* (which we couldn't help but think of as man-dresses)—that it was difficult to distinguish one from the other. Or, for that matter, to tell them from the general population. They were indeed, as Mao Tse-tung so famously put it, like fishes in the sea.

Whatever intelligence I and the other Nellis Predator crews picked up was passed through channels for action. For several days we watched armed guerrillas coming and going from a mid-city hideout. Finally, somebody in higher headBobbys made a decision to strike. I observed from above while a marine F-18 fast mover hit the house with a smart bomb. About ten to twelve Tawid operators were gathered inside. A dirty brown mushroom erupted and spread smoke, dust, and debris over the entire block. None of the terrorists escaped unscathed; Zarqawi alas was not among the casualties.

Fallout from the bombing was minimal compared to the fallout from around the world after Arab media outlets got through with the story. Zarqawi and his ilk knew how to work the media. It was a common view among U.S. soldiers that Al-Jazeera was sympathetic to the insurgents. Its reporters were virtually the only ones allowed access to insurgents without fear of being harmed.

According to the network, witnesses denied that the house was hiding militants. "We have nothing to do with the resistance of al-Zarqawi," insisted the owner of the house, who was absent from home when the bombing occurred. "These are pretexts of the U.S. military to terrorize the people in Fallujah because U.S. soldiers are unable to face the insurgents."

Other Iraqi witnesses claimed that a market, other houses, and an ambulance were also bombed. I knew better, of course, as I had watched the entire episode. The ambulance was supposedly struck while rushing from the area with five wounded patients; the paramedic and all five patients were allegedly killed.

Al-Jazeera aired footage of the hospital in Fallujah overwhelmed with wounded, its white sheets splattered with innocent blood. One woman screamed hysterically and ripped out her hair. "I lost my son," she sobbed. "I wish it were me."

There had been no similar histrionics in the aftermath of Zarqawi's having coldly beheaded his bound and blindfolded victims for the world to see.

An enormous rundown Shiite mosque occupied a plot of more than a hundred acres of land on the western edge of Baghdad's city limits. Saddam Hussein was a Sunni. Under his rule, Shiites had been persecuted to the point that many Shiite mosques had enjoyed little use in recent years. Weeds had overgrown narrow roads leading to the shrine, all of which were potholed and littered with road fill and trash. Guerrillas sometimes used the brush-choked grounds from which to lob mortar rounds at patrolling U.S. troops.

Balad Air Base, on the outskirts of Baghdad, hangared our MQ-1s and housed the LRE unit that launched and recovered them for us. One morning, minutes after takeoff and my assuming command of the aircraft, my chatroom erupted with excited conversation. An army patrol had just taken a mortar round from the vicinity of the old mosque.

Speck Three-Two, we've had a contact at the mosque. See if you can pinpoint the SOB.

"Speck" was the regular call sign for this daily Predator mission; only the numbers changed every day. Darknight was the on-ground JTAC.

Roger that, Darknight, I typed back. *I'm south. ETA in five mikes.*

He's yours, Speck, came the response on the screen, *if you can root him out.*

That meant a possible kill for my logbook.

United States forces had enjoyed air superiority almost from the day the war started. In fact, very few of Saddam's warplanes ever got off the ground. Since I didn't have to worry about some pesky MiG showing up to shoot me down, I could concentrate my full attention—and my cameras—on the mosque and surrounding terrain. I slowly circled, searching for movement in the brush almost inch by inch.

I was still hunting when a man equipped with a broom and a rusty shovel with a broken handle appeared out of nowhere and began working on one of the dilapidated roads that branched off the thoroughfare to the mosque. At least that was what he appeared to be doing. You could never be sure, however. Someone digging in a road was always cause for suspicion. He could just as well be a saboteur planting a bomb. Improvised explosive devices accounted for more dead Americans than any other weapon in the insurgents' arsenal.

My crew and I watched the guy closely for some time. He had no clue whatsoever of my presence in the sky. In fact, he didn't seem to care if he were noticed or not. He straightened and arched his back, stretching. He yawned. He wiped sweat from his brow. He appeared to be in his thirties and, best I could tell, had very little spare flesh on his rib cage. I finally concluded that he was exactly who he seemed to be—some ordinary Iraqi citizen clearing debris off the road and filling in potholes, perhaps tidying up in hopes that the unused mosque might be revitalized.

By now, whoever ambushed the patrol was long gone. Unharmed, the patrol had also moved out. I typed my conclusions about the lone Iraqi into the chatroom with the army JTAC on the ground. He wasn't satisfied.

We're sending out another patrol to check him out, Speck. Will you monitor?

It wasn't long before a cavalcade of Bradley armored infantry vehicles crawled into view on my screen. The vehicles pulled up to the man and stopped. The Iraqi leaned on his broom and waited. Some of the Bradleys assumed a tactical defense with machine guns pointing outboard like the horns of so many threatened African cape buffalo.

The Iraqi slowly shuffled to the shoulder of the road and leaned on his broom again. A U.S. officer who was perched in the turret of the lead armored car and the road worker stared at each other for several more long moments.

Speck, he seems innocent enough.

Roger, Darknight. Maybe you should give him a medal for good citizenship.

Negative. He's probably the brother-in-law of the guy we're looking for. LOL.

A smiley face made from a semi-colon and a parenthesis popped up on my screen to reinforce the "LOL" (laughing out loud), and the armored cars began moving on. Out of carelessness or thoughtlessness, I assumed, certainly not from sheer maliciousness, the convoy drove over the section of road that the Iraqi had already repaired or cleaned up. Each of the vehicles weighed about twenty tons. They further cracked the macadam surface and reopened potholes, destroying all the progress the worker had made.

He stood with his broom and broken shovel by the side of the road and gazed dejectedly at the armored convoy as it drove off and over the horizon. His shoulders slumped and his chin dropped onto his chest. I recognized the posture even from ten thousand feet above him. Defeat. Poor bastard. It must seem that every time he rolled up his sleeves to get about the hard work of rebuilding a life, someone came along and ruined it.

Some scenes of war were smaller than others while their impact was greater.

"To a soldier in a hole," cartoonist Bill Mauldin wrote in his World War II classic *Up Front*, "nothing is bigger or more vital to him than the war which is going on in the immediate vicinity of his hole. If nothing is happening to him, and he is able to relax that day, then it is a good war, no matter what is going on elsewhere. But if things are rough, and he is sweating out a mortar barrage, and his best friend is killed on patrol, then it is a rough war."

Mauldin's observation could have applied to the Civil War, the war in Iraq, or any other war just as readily as to World War II. Individual soldiers throughout history have owned only a small piece of their war. Few have been able to see the big picture in order to understand larger events that influence the ebb and flow of the fighting. Flying the Predator allowed me the extraordinary perspective of being not only a "combatant," albeit from 7,500 miles away, but also an observer with a broad overview.

I saw the war firsthand day after day as it unfolded. It seemed I was always watching in real time, hovering above, sometimes swooping down to join it like the predator I was. At the same time, I was "the Professor," an amateur student of history. Unlike the common grunt on the ground, I was a ten- to twelve-hour-a-day warrior. I had time to read books and magazines and newspaper articles, watch TV, or take Trish out to a movie.

Thus, I knew when the president of the United States met with NATO to ask kindly for more foreign participation in the war. I understood the various missions and operations devised by on-site commanders. I viewed antiwar protests on TV and saw captives of al Qaeda beheaded on the Web (although I could never get myself to watch the actual beheading). Columnists and pundits, news commentators, clergymen, and probably every politician who could find a camera expressed an opinion. Everybody had one, as the saying went, and they all stunk.

Other than Baghdad, the two cities in Iraq claiming the most ink and airtime were Fallujah, where Abu Musab al-Zarqawi reigned, at least temporarily, and Najaf, where Muqtada al-Sadr, the cleric and titular head of the Mahdi Army, was still holed up in the Imam Ali Mosque. The entire world seemed intrigued by the moon-faced mystic, who reminded me of Chucky the terror doll in the movies. Ringed in by marine steel, al-Sadr and his fighters and leaders defied the world's mightiest army. Maybe we weren't going in to root them out, but they weren't going anyplace either. The Crusaders and the armies of Saladin had laid siege to each other's cities in much the same way. Sooner or later, I assumed, we would get fed up with al-Sadr's antics.

With the Predator, as well as in the media, I followed the Mosque siege as it unfolded day by day, as captivated by the situation and al-Sadr as the rest of the world. It was almost like watching some reality TV program that went on endlessly. It had to end at some point.

Now thirty-one years old, Muqtada was the fourth son of the famous Iraqi Shiite cleric Grand Ayatollah Mohammad Sadeq al-Sadr, who was murdered by the government of Saddam Hussein and in whose honor the slums of Sadr City were named. Due at least partly to his father's prominence, the younger al-Sadr rose quickly to become one of the most influential religious and political figures in the country not holding an official title in government. He proved himself as ruthless as Chucky the terror doll when it came to those who opposed him.

In April 2003, a month after the start of the war, al-Sadr and his followers used the breakdown in civil society to seek revenge against Imam Haidar Raifee, who al-Sadr accused of fostering the conspiracy to assassinate his father. Both al-Sadr and Raifee claimed control of the Imam Ali Mosque. A third imam, Abdul Majid al-Khoei, tried to broker a settlement between the rivals by arranging a meeting between them. An angry crowd inspired by al-Sadr confronted al-Khoei and Haidar Raifee when they arrived at the shrine. The mob killed Raifee with bayonets and knives, then dragged al-Khoei before al-Sadr.

"What shall we do with him?" the mob chanted.

To which al-Sadr might have coldly responded, "Take this person away and kill him."

For the next year, from April 2003 to March 2004, Muqtada al-Sadr, with the backing of his Mahdi Army and his "Sadr Bureau," ran Sadr City unopposed—until residents decided to elect neighborhood councils and ultimately a district council to represent them in Baghdad. Al-Sadr was unwilling to give up power. The Mahdi Army revolted against the councils and American influence on April 4 by ambushing a U.S. Army patrol, killing eight Americans and wounding fifty-seven more.

I began flying Predator over Iraq shortly after two months of bitter fighting in Najaf and Sadr City between coalition forces and the Mahdis left both sides exhausted and in an unstable truce. On July 31, al-Sadr broke the ceasefire after U.S. Marines and the Iraqi National Guard raided a safe house in Karbala and nabbed some al-Sadr representatives. Al-Sadr issued a blatant challenge to the new government, demanding that his people "be freed, and if this is ignored then we will respond at the appropriate time."

Iraqi police and U.S. troops surrounded al-Sadr's house on August 3 and engaged in a furious firefight with hundreds of Mahdi fighters defending the house. Clashes spread to the old city of Najaf. By August 13, the cleric and the main body of the resistance were trapped inside a cordon around the Imam Ali Mosque. Day after day I flew over the shining dome and its twin minarets and watched insurgents below brazenly shooting rockets and mortar rounds indiscriminately into the surrounding neighborhoods.

It looked like the stalemate might finally reach a conclusion as August drew toward an end, thanks to Grand Ayatollah Ali al-Sistani. At seventy-four years old, he was an Iranian and "Twelver" (those who believed that the return of the Twelfth Mahdi and the end of the world were imminent) who had resided in Iraq since 1951. He returned from London, where he had sought medical treatment, and traveled to Najaf in a "peace convoy . . . to stop the bloodshed." Al-Sadr was apparently ready to call another truce; the Mahdi resistance had suffered hundreds of casualties since April, whereas U.S. Marine losses were fairly light.

The following day, al-Sistani announced that he had compromised an agreement with al-Sadr: The Mahdi Army would voluntarily disarm and leave Najaf if U.S. forces withdrew from the city and returned control of it to Iraqi authorities. I watched from the air as the disarmament process unfolded. Iraqi police in their blue-and-white Renaults hauled away loads of RPGs, mortars, and rocket launchers and piles of AK-47s. So many weapons stockpiled in one place was a disturbing sight, underlying how easy it was to smuggle foreign weapons and foreign fighters into the country.

Hopefully, the Iraqi government made sure the weapons were destroyed—although I had my doubts. I didn't much trust the Iraqi police, not after what I had witnessed of them in Fallujah.

It was a strange war. We had had all these enemy fighters cornered. At any other time, in any other place, we would have cut them off and killed them. Or at least taken them prisoner. Now, they streamed out of the mosque in broad sunshine, presently unarmed to be sure, and melted into the city and countryside. I guessed it wouldn't take them long to rearm and come back for another round.

CHAPTER 8

Are We There Yet?

THE U.S. OPERATIONAL APPROACH at the time held that the best chance we had of winning the war in the city streets of Iraq was to capture or kill insurgents faster than they could replenish themselves either from homegrown sources or from foreign infiltrators. We therefore pressed that approach to its logical outcome. As time went on, the insurgents began to understand that they couldn't conduct their plotting and other nefarious activities right out in the open or we would find them and kill them, although they didn't understand how we did it. They thought they were being so clever about covering their tracks that they were invisible, like Mao's fish in the sea. The poor bastards never once considered looking up, *way* up, from which heights Predator crews observed their every move, where they went and who they met with. We were always hunting, day and night.

For a week I had been watching a house in Fallujah known to be frequented by guerrillas and terrorists. Sooner or later, I was bound to cross trails with Abu Musab al-Zarqawi and put an end to his reign of terror. He couldn't hide out forever. As I panned cameras across the target house and into the neighborhood, I noticed several men acting suspiciously in the parking lot of a little greasy spoon café across the street. Defining *suspicious* wasn't always easy. It was like a cop who had a sixth sense that somebody or something was out of place, a sense that allowed him to distinguish criminal activity from the normal day-to-day routine.

As I turned back for another look, the men began loading boxes into the trunk of a faded-red compact car. I couldn't tell what they were handling, but I doubted it was a shipment of olives. They finished what they were doing. All but the driver went back into the café. The driver slammed the trunk lid and looked all around. That was a dead giveaway for *suspicious*. People with nothing to hide didn't care if other people were watching them. I decided to follow the car when it pulled into city traffic.

It snaked across town and entered a courtyard. Rather than lawns, most houses in Iraq had courtyards of bare, hardened earth that were enclosed by mud-brick walls. The wide branches of a eucalyptus obscured much of the bustle and excitement that greeted the car's arrival, but what I could see was enough to convince me that this was probably not the home of some local law-abiding baker.

A large moving van was already parked inside the enclosure next to the gate. Men were rushing in and out of the house, but I still couldn't tell what they were up to. Soon, another vehicle, a dark Peugeot, pulled in for a short visit. I recognized it from my "BOLO sheet" as belonging to one of al-Zarqawi's men. I took a sip of coffee, stretched my muscles, and waited, in the meantime tagging the house. The analysts would later name it *El Campo* (we had a Spanish theme going that week). Our maps of Baghdad and other cities were dotted with code-named sites for ready reference. For example, we called the first house across from the café *Berlin*. The café was *Richardson*. This was our first encounter with *El Campo*; it appeared to be a target-rich environment.

Zarqawi's man left *El Campo*. I decided to pull off the original dull red compact in favor of tailing this guy in the Peugeot.

"He's up to something," I said to Staff Sergeant McKenzie, the sensor occupying the seat next to me.

"Aren't they always?"

"Only when they're awake."

McKenzie and I followed the car all over Fallujah for the next two hours. Good intelligence work required the patience of a Zen master. The guy stopped off at several places to make deliveries—of what I still couldn't determine. Everything was in boxes or covered in some other manner. I suspected ammo. Then he headed out of the city at a good clip.

Whatever he was hauling, he obviously didn't want his vehicle searched. The U.S. Marines were still barred from entering Fallujah, but they continued to cordon off the city with roadblocks and checkpoints on the outskirts. Each time my target approached one of them, he pulled a quick U-turn and looked for an alternate route. This was part of what I meant by *suspicious*.

He soon selected a back road out of town that the marines weren't covering. He exited the city and drove around the countryside for a while, making fast turns and cutbacks to check his back trail. Finally, he whipped down a dirt road that led to an insurgent safe house code-named *Disney World*.

"I knew it!" I exclaimed and immediately transferred my video feed to marines on the ground so they could see what I was seeing. The guy was fair game the minute he crossed out of city limits.

My chatroom started chattering.

Keep on him, Speck. We're on the way. Can you push to Warhorse One-One on White One-Two-Eight?

Roger.

Warhorse One-One was the forward air controller imbedded with a pair of Blackhawk helicopters filled with a marine quick reaction force. White One-Two-Eight was his coded radio frequency. I switched channels to voice.

"*We're Mary in two mikes,*" Warhorse radioed. That meant they would be airborne within two minutes. Marines wanted this guy now in the worst way.

"*This is Speck Three-Two. I got him still heading west. Possible weapons loading.*"

"We got your video. Looks like the dark Peugeot we've seen before."

Operational chatter picked up a notch when the choppers streaked across the screen below my aircraft and acquired their own visual of the target. So far, the guy hadn't done enough to warrant our blowing him off the road, but we certainly had sufficient probable cause to capture his scheming hide.

He must have soiled himself when he heard the roar of the helicopters and looked up. He gunned his engine to escape. The pursuit was down dusty roads. One of the Blackhawks swooped low over the car. The chopper surged ahead, then flipped around and dropped down on the road, blocking it. The other helicopter set down behind the car to cut off retreat. It was like the desert version of *Cops*.

Without so much as a moment's hesitation, the guy twisted his wheel. The dark Peugeot swerved off the road in an explosion of dust, sand, and gravel and tore across the flats at full speed. The choppers clawed back into the air and gave chase, skimming the nape of the earth and easily keeping pace. Warhorse inside the lead chopper jumped on the radio to coordinate a trap with an M1A1 Abrams tank that was already racing at forty miles per hour to intercept.

From my unique vantage point, I watched the entire drama play itself out in real time: the fleeing bad guy chased by two Blackhawks, the tank etching a long billow of dust in its wake as it closed in. The terrorist had no idea that the tank was anywhere in the vicinity until it suddenly loomed out of a shallow wadi and charged headlong at him. He almost rolled his vehicle trying to avoid colliding with the steel behemoth, finally stalling his car in a roiling cloud of dust. I held my breath when the Abrams thundered directly at the car and pointed its 120mm main cannon right at the driver through his windshield. Talk about commanding respect.

The choppers hovered. Marines spilled out armed and prepared for a fight. Apparently the guy wasn't ready for his seventy-two virgins in Paradise. He piled out of the car with his hands in the air. Marines cuffed him and stuffed him into one of the Blackhawks while other marines searched the vehicle. I later learned that our hero had indeed been delivering ammunition to various insurgency strongholds.

Everything was just winding down when my relief pilot sauntered into the GCS. I rose from my seat to let him take over the cockpit, still chuckling from the excitement of the chase and its zany climax.

"What's so funny?" Maj. "Slack" Roberts asked. "How about taking my shift if you're having such a good time?"

I slapped him on the shoulder. "I've had all the fun I can stand for one day. I'm taking Trish out tonight."

He sat down. "War is hell," he observed.

It was enough to make a Predator pilot schizophrenic, what with fighting two wars simultaneously 1,500 miles apart and balancing them with a wife and kids, if he had them, paying the bills, and calling the plumber because the toilet was stopped up. It didn't get much more surreal than that.

"Honey, you seem a million miles away," Trish would notice.

"Sorry. Not quite that far away. Sometimes it's hard to keep switching on and off. Back and forth. It's like living in two places at the same time. Parallel universes."

As much as I would miss her and home, I thought it might be easier when it came my turn to deploy to Iraq toward the end of the year to work the launch and recovery element. Landing and taking off airplanes for pilots back at Nellis. In-country *with* the war.

In the meantime, the United States continued to increase its military presence in Afghanistan, which meant requirements for additional Predator hours flying above the Hindu Kush Mountains and along the Pakistan border, across which Taliban and al Qaeda fighters flowed almost at will while we were restricted to the Afghan side. Back in the nineteenth century, Mexican bandits used to operate the same way, crossing the Rio Grande to raid American settlers, then hightailing it back across the river. In fact, the first use of American warplanes in a foreign country happened in 1916 when the 1st Aero Squadron chased Poncho Villa around the West Texas and Mexican countryside. I often felt like we were doing the same job, almost one hundred years later, with much better technology but not much more success.

Compared to Iraq, flying Afghanistan was like taking a long, slow trip cross-country. *Are we there yet?* One morning, one of the units we supported asked me to relocate from "Nixon's Nose," on the Pakistani border, to the town of Herat, on the Iranian border—a five-hour flight with nothing to see except brown valleys and brown mountains topped with snow. Our intel people thought that Taliban insurgents intended to disrupt the city in order to undermine confidence in the Kabul government. Personally, I doubted the Taliban would be welcome there.

Herat was the most modern and prosperous city in Afghanistan. The Soviets failed to reach Herat, so it escaped being bombed into the ground during that long war. After the Russians left, the local governor managed the city as if it were an autonomous city-state. It remained virtually the only city in the country where the electricity worked, the government functioned, cops walked the beat, schools were still open, and paved roads had no potholes. There were even regular elections. Why would a city like that *want* the Taliban?

I settled back for the flight, scanning my cameras to ease the monotony.

Afghanistan was an old, old country. I felt its antiquity as I gazed down upon terrain that resembled a sheet of paper crumpled into valleys and mountains by some giant fist millions of years ago. Archaeologists estimated that humans had been living here for at least the past fifty thousand years. Its farming communities were among the earliest in the world—and many regions seemingly had not modernized much since then. Wheelbarrows, bicycles, and donkeys were more common than Toyotas and Renault trucks in many places. Sometimes a donkey was rigged with an IED to become a weapon instead of a beast of burden.

Landlocked and largely mountainous (Nowshak at 24,557 feet was one of the tallest peaks in the world, only a couple of thousand feet lower than Mount Everest), the country was bordered by Pakistan, China, and Iran, with Tajikistan, Turkmenistan, and Uzbekistan to the north. Looking down on to that ancient focal point between East and West connected by the old Silk Road, I could almost feel the presence of Alexander the Great and Genghis Khan's Mongols. And Rudyard Kipling. *"By the livin' Gawd that made you, You're a better man than I am, Gunga Din!"*

There was an old mistruth often quoted to the effect that Afghanistan was unconquerable. Ungovernable perhaps. But it had been the home of many conquerors—Aryans, Median and Persian empires, Greeks, Seleucides, Turks, Mongols, and English.

In 1219, the Mongols under Genghis Khan devastated the land, as they did later in Iraq, exterminating virtually every human being in the ancient cities of Herat and Balkh. Destruction caused by the hordes depopulated other major cities and frightened much of the population back into an agrarian rural society, from which, like other countries overrun by Mongols, Afghanistan never completely recovered. It was also during these Middle Ages that Islam established roots in the country.

Modern Afghanistan began to emerge in 1747 when Ahmad Shah Durrani created the Durrani Empire. The British took over a bit more than a century later, turning the country into a buffer state in "The Great Game" played out between the British Indian Empire and the Russian Empire. Afghanistan gained its independence in 1919 following the third Anglo-Afghan war during which the English were expelled. It was out of these wars that the myth of "unconquerability" grew.

Afghanistan since the late 1970s has known almost nothing but civil war, marked by occasional foreign invasions. The Soviet adventure that began in 1979 may have killed as many as two million civilians before Russia withdrew with its tail tucked between its legs ten years later. The United States had armed and supported the Afghan Mujahidin. World players in terrorism such as Osama bin Laden and Abu Musab al-Zarqawi emerged from the Russian years to ply their bloody trade against the West in general.

Fighting raged among the various Mujahidin factions after the Soviet Union left until, by 2000, the Taliban had captured most of the country and established the Islamic Emirate of Afghanistan. Following 9-11, the United States and its allies launched Operation Enduring Freedom and invaded the restless and war-torn country to destroy al Qaeda training camps and capture Osama bin Laden and his lieutenants. Teams of CIA paramilitary officers and U.S. Army Special Forces helped the Northern Alliance overthrow the Taliban and seize Kabul, the capital. Hamid Karzai would become the nation's first elected president, ever.

That didn't mean the fighting was over in Afghanistan. Guerrillas sought refuge along the Pakistan border, in isolated villages, and in the mountains where they continued their stubborn resistance in behalf of the Taliban and al Qaeda. That was where I came in, flying Predators over country more accustomed to scruffy little Mongol ponies than twenty-first-century technology. Amazing. I was merely the latest in a long line of invaders—and the process wasn't likely to end with me.

I arrived over Herat, made one pass, and saw nothing unusual unless you counted the unnatural condition of peace as out of place.

Pacman One-Six, you are requested to return to Kabul ASAP.

Modern warfare was as much about public relations as it was about shooting. If we bombed a terrorist's parked truck and got some shrapnel in an abandoned ambulance next to it, some media outlets were all too willing to go along with insurgents' claims that Americans had dropped ordnance on an ambulance full of women and children. The same thing happened if we caught a bunch of Muslim-killing foreign fighters in a safe house; miraculously, they morphed into innocent cobblers and goat farmers and loving fathers by the time the world's media were finished with it. Every time we released a prisoner nabbed at an IED site or fleeing the scene of a sniper attack, generally because we had neither the time, the inclination, nor the evidence required by the new Iraqi legal code to *know* that he was the enemy, the first thing the guy did was run to the press with tales of atrocities committed against him. Such as being forced to listen to loud ZZ Top or Michael Jackson music while hanging upside down by his toenails (only the first part of which we might be willing to do).

Of course we had made strategic mistakes—Abu Ghraib being chief among them. The techniques that U.S. policy had officially sanctioned as forms of "enhanced interrogation" had emboldened a small group of under-supervised military policy to commit war crimes against Iraqi detainees and capture it on film. And to this day the image of the United States as liberator and vanguard of justice is tarnished.

It therefore wasn't a surprise to me that U.S. action was bound to be mistrusted by the average Iraqi citizens. But one afternoon while I was patrolling up north in Iraq's Kurdish country, I saw for the first time evidence that the insurgency was running its own "hearts and minds" campaign. I noticed a pickup driving around Tell Afar with two men in back handing out boxes. It appeared they were distributing emergency food packets, except our senior intel coordinator assured me that multinational forces had no aid workers in the city. And as for the fledgling Iraqi government, it was barely able to set out an agenda for its next meeting much less organize an aid distribution effort.

That left only one possibility—the insurgency network. The anti-Israel terrorist organization Hamas had done this same thing in Palestine. In spite of a long record of rocketing Jewish schools and pizza parlors, Hamas turned itself into good guys in one fell swoop of media ink by distributing food, building schools and clinics, and patrolling the streets in Palestinian neighborhoods.

It therefore became personal with the Palestinian people whenever Israel retaliated against terrorist acts. The world press expressed outrage. After all, Hamas was a community betterment organization, right? And maybe they were.

I knew that the same scheme was under way in Tell Afar, but there was nothing we could do about it. The insurgency was busy winning hearts and minds while the Iraqi government was too weak to secure the area of operations (AO) and allow aid from the UN and other relief organizations to safely enter the area. And maybe that wasn't the worst thing in the world if some good came of it. But it was a win-win proposition for the insurgency. Imagine world opinion if we planted a Hellfire in the back of a pickup distributing box lunches for hungry Kurds. No matter that underneath the food were likely loads of AK-47s, mortar tubes, and IED materials that would wind up killing as many innocent civilians as coalition soldiers.

I had to hand it to these particular insurgents, however. They were clever.

CHAPTER 9

Never Enough

IN SPITE OF CONCESSIONS and despite the so-called truce, Fallujah remained a hotbed of insurgent activity spurred on by Zarqawi and his Tawid guys. Sooner or later, marines would have to go back in. The Iraqi police had already proved they couldn't handle it alone. Even those cops who were honest and upright were swimming against a tsunami. They had the most dangerous job in the country.

Americans helped harden police stations to make them less vulnerable to truck bombs and suicide martyrs. Insurgents responded by obtaining the names of police recruits and assassinating them in their homes, on the streets, or at recruiting depots. Police beefed up security in those areas. Terrorists then took to bombing shops and cafés known to be frequented by police.

Whereas Americans wore body armor and drove around in armored Humvees, Abrams tanks, and Bradley fighting vehicles, the Iraq police sported about in Renault hatchbacks and Toyota pickups—and were wiped out almost every time they encountered an ambush. In cities other than Fallujah, that led to U.S. armored escort for police on patrol as well as armed aerial surveillance of police stations. Playing one-upmanship with blood. In Fallujah, where the police were on their own, their numbers dwindled due to attrition, treachery, and fear of enlisting, all of which contributed to even greater disorder and more anarchy. There would therefore have to be a reckoning when the marines returned.

One evening when I reported to the POC for my shift, I discovered a commotion around the MCC's station. Off-duty pilots, sensors, maintainers, and anyone else not actively engaged were grouped around a plasma screen that showed some insurgents blatantly driving around with a .50-caliber machine gun mounted in the bed of a pickup truck. They were obviously trolling for trouble, their gunner ready for action. Stupidity knew no limits. I assumed this had to be Fallujah.

Captain Tom Conroy, the pilot I was scheduled to relieve, chose to remain in the cockpit. It was his chance for a little action. And for the sake of tactical continuity and uninterrupted situational awareness, it was squadron policy to keep a crew in the seat when things heated up. I remained with everyone else as spectators at the MCC plasma screen.

Predator tailed the truck while Captain Conroy waited for it to stop somewhere to provide him with a stationary target. It finally pulled into a courtyard and backed into a residential garage. The early-morning sun in Fallujah shone on the garage door as it closed.

Ops officers, pilots, the MCC, and the sensor operators engaged in a spirited discussion over the best method for striking to destroy the truck and the machine gun and its crew. Of course, the final decision over what action to take wasn't ours to make. The ground commander had the last word. He opted to hit the house with a fighter-dropped 500-pound JDAM, which was a lot more powerful than a Hellfire and would leave little more than a smoking hole in the ground where the house now stood. The commander ordered Captain Conroy and his crew to stand by and keep an eye on the target to provide battle damage assessment (BDA).

Tension mounted at the POC in anticipation of watching the house demolished. A JDAM-armed F-16 Fighting Falcon (in the Air Force we call them "Vipers") was on its way. Estimated time of arrival (ETA) in three minutes. However, the situation changed abruptly when the garage door opened and the truck and its ragtag band of insurgents sped off, the gunner spread-legged with his head thrown back like he owned the city, fingers gripping the handles of the mounted machine gun. This bunch was obviously on the prod. They were probably scouting for rival gangs to shoot up because there were no Americans in the city.

The JDAM that the F-16 carried came equipped with a guidance kit that converted an unguided gravity "dumb" bomb into a "smart" bomb. An integrated inertial guidance system coupled with a GPS receiver directed the ordnance to its target. It was programmed to fly to a particular designated point and then detonate. It could strike a set of coordinates with great accuracy, but it couldn't hit a moving target.

The Toyota paraded around town for about twenty minutes, making a show and coming across no other technicals or gangs that posed a threat. Even the police seemed to be steering clear of it. By then, the F-16 was running low on fuel, and it returned to base. Another wouldn't be available for at least an hour. That meant Captain Conroy and his Predator were moved to center stage. I left the POC and hurried to the GCS trailer to watch Conroy take the shot if one presented itself.

The Toyota darted into another garage at the edge of town. But while the command centers in Iraq and Qatar were trying to figure out a plan, the truck took off again.

"This is ridiculous!" I exploded. "What are they waiting for? Those guys are going to get away."

The truck with its raucous load of gunmen proceeded to another neighborhood near the warehouse district and backed into still a third garage. It was as though they were trying to find one that fit, or somebody who wanted them. This time the garage door apparently malfunctioned and only closed about a Bobby of the way down. The front half of the Toyota remained visible on the HUD. A message from the ground commander flashed across Conroy's chatroom.

Agony Three-Zero, target is declared hostile, start your target run. Expect clearance on final.

About time.

Tom glanced up at me. I bent over his shoulder to offer a word of encouragement. This was going to be a difficult job. The Special K he carried would arrive on-target at a forty-five-degree angle, which meant it might not penetrate the garage far enough to score a successful hit.

"Place the crosshairs just below the door," I advised. "Then light up the fires."

We dared not wait any longer. The truck would probably be moving again shortly, if its past history was an example.

Conroy and his sensor initiated their attack run. Through the crosshairs on the video screen, I saw the sensor's laser beam planted high on the front edge of the pickup's hood, just below the leading edge of the partly open garage door. The missile had to "see" and track the laser or it would run off course and end up nobody knew where. He received clearance from the JTAC.

Captain Controy "pickled." The missile was on its way. Seconds passed. I held my breath, along with Tom and his sensor.

A bright explosion washed out the screen. It looked like a good shot—until the smoke and dust settled, after which it became clear that the machine gun had escaped unscathed. Only the front of the truck appeared damaged. We might have been more successful had we punched an M Model through the roof.

It was too late now. As always, civilians began swarming all over the place. Even though they had no idea where the missile came from, they knew we wouldn't fire a second and chance hitting innocent people.

I suppose you could have called it progress vis-à-vis the War on Terror. The United States had just taken off the front of an old Toyota at the price of about sixty-eight thousand dollars. Eventually we might get a handle on the insurgency—if we didn't go broke first.

***

Abu Musab al-Zarqawi was at it again in Fallujah. The murderous, cold-blooded terrorist specialized in personally beheading his hostages and showing it on TV and the Web. Nicholas Berg hadn't been Zarqawi's first in May 2004, nor would Berg be his last. Four months later, the al Qaeda–appointed "Emir of al Qaeda in the Country of the Two Rivers" and his brutish henchmen seized two American contractors, Eugene Armstrong and Jack Hensley, and a Briton named Kenneth Bigley and promptly declared that he would behead them unless the United States freed all women prisoners being held in Iraqi jails. He had to mean "Mrs. Anthrax" and "Doctor Germ," the only two females detained in Iraq, both of whom were charged with working on Saddam Hussein's biochemical weapons program.

Zarqawi made good on his threat. Only days after the abductions, a videotape airing on a terrorist website showed the terrorist leader executing Armstrong. The production opened as usual with the victim bound and blindfolded and on his knees in front of the camera; behind him in a semicircle was a group of masked men all in black. The man intel identified as Zarqawi stepped forward and snatched back the victim's head to expose his bare throat. Blood spurted and Armstrong emitted wet, strangling sobs as Zarqawi sawed off his head with a large and apparently dull knife. The murderer then held up the severed head and shoved it toward the camera for a closeup.

"*Allahu Akbar!*" the executioners shouted. *God is great!*

The clip ended with Zarqawi vowing that more heads were going to fall.

Given the opportunity, I would blow him to hell and then take Trish out to dinner and think nothing else about it. The monster had now murdered another American and still held a second and the Brit in the wings. Everyone knew they were done for if we didn't pull off something fast.

In my opinion, our best option for finding the surviving hostages lay in the busy market district of downtown Fallujah, which seemed to be a popular hangout for Zarqawi's insurgents.

My crew and I soon picked up on a man thought to be a Zarqawi lieutenant as he wended his way through open stalls selling everything from skinned goats hanging by their heels to the latest Islamic anti-American videos. I hoped he might lead me to either Zarqawi or the place where Hensley and Bigley were being held.

He entered a one-story flat-roofed building with a small vendor's sign hanging over the door. I couldn't quite make out the lettering, which would not have helped me much anyhow because it was in Arabic. I guessed it was probably a law office.

I settled in to wait for him to come out or for something else to happen that might provide a clue. Circling at eight thousand feet or so, all but invisible, I wasn't too concerned about being seen. The Predator, with its powerful high-resolution telephoto lenses, was one remarkable piece of equipment. Not to mention the twin Hellfires strapped beneath its wings.

As I watched, a crowd began to gather at a nearby intersection. I asked my SO to zoom in for a tighter picture. There was a lot of vigorous discussion and gesticulating going on down there. I knew that something was up when my guy rushed from the building where we had left him and jumped into a black car already occupied by three other men. The crowd immediately dispersed to other cars and followed the first.

Curiosity whetted, I followed. Other vehicles pulled out from side streets and driveways and alleys to join the procession until it was nearly a mile long.

"I love a parade," Senior Airman Mata sang softly from his sensor's console.

We shadowed the motorcade to what at first appeared to be an empty field near the edge of Fallujah but which was still within the cordon drawn around the city by U.S. Marines. More than a hundred cars, trucks, and motorbikes arrived and parked willy-nilly all over the place. For all I knew, they were going to have a rally, a protest, or something.

To my surprise, some men unloaded a casket from an SUV and carried it to a freshly opened grave. A funeral! But not just a funeral. To my even greater astonishment, the mourners produced sixteen more caskets, some of which were sized for children. There appeared to be little formality as the crowd separated into groups around individual graves and seemed to compete for being the most bereaved.

As conditions continued to deteriorate in Fallujah, marines and air force fast movers had stepped up air strikes on insurgent positions. The U.S. forces had been accused in the media of indiscriminate killing in Fallujah. But there was no way our recent air strikes (assuming they were observing the Muslim custom of a quick burial) could have accounted for so many bodies. The air strikes over the last few days had been quite limited, against isolated targets. Most of the departed must have died from other causes such as disease, starvation, or murder. There was almost no government of consequence, and the city was full of competing groups—Sunnis versus Shiites; native Iraqi fighters versus foreigners; locals versus strangers; one section of town against the other. Shootouts, assassinations, and other random acts of violence were common. The city reminded me of the *Mad Max* movies in which survivors of a worldwide holocaust fought one another for dominance, power, and dwindling resources.

I couldn't help harboring empathy for all that sorrow concentrated in one place down there. It was almost as if the townspeople were gathering to bury the city itself.

My guy soon departed the funerals in the black car with his three cronies. We pulled off the cemetery to keep on his tail. Intermission was over. If we could locate and kill Zarqawi, not only might we save the captured contractors but, at the same time, we could prevent other mass funerals. The bloodshed would never cease in Fallujah until we caught or killed the terrorists who incited it.

The black car returned my guy to the law office, where he remained as night settled in. Patience was one of a Predator pilot's finest attributes—but I couldn't help feeling antsy. Time was running out for Hensley and Bigley.

***

In its Web posting the night after Zarqawi murdered Eugene Armstrong, Tawid and Jihad issued a chill warning to President George W. Bush and British Prime Minister Tony Blair. The video assured them that Hensley and Bigley would be relieved of their heads within twenty-four hours unless the two nations capitulated to terrorist demands—demands that were essentially impossible to meet even had the two allied nations been so inclined. The Western press informed Zarqawi that no insurgency women were being detained.

A voice-over that accompanied the posting summarized in a bitter, anonymous monotone: "The Muslim blood is not water, and the honor of Muslim women won't go to waste. Bush, eat your heart out, and Blair, may you cry tears of blood. God is great. Glory be to Him, His prophet and the faithful."

President Bush issued a statement declaring that U.S. policy did not include negotiating with terrorists.

"You cannot negotiate with these people," he said. "They will behead people in order to shake our will. These people are ideologues of hatred. We will stay on the offensive against them."

Zarqawi put Kenneth Bigley the Brit on the Web. The video showed him teary and begging for help.

"Please, please help me. I need you to help me, Mr. Blair . . . I think this is possibly my last chance to speak to somebody who will listen. I don't want to die here."

As far as we knew, both hostages were still alive when I took over a GCS for my shift, relieving Capt. Jesse Winters and his sensor. I had barely settled into my seat at the controls when I received a new mission through the chatroom. It blinked red, indicating a high-priority message. A human intelligence (HUMINT) source on the ground supposedly *might* know where the contractors were being held, not in Fallujah as we first thought but in Baghdad not far from the Green Zone. They needed an MQ-1 to take a look. I poured on the coal and barreled at a whopping 110 knots (about 125 miles per hour) out of Fallujah heading for central Baghdad.

From the air, when we arrived, I commanded an unobstructed view of the middle of Baghdad on the northern shore of the Tigris and the ring of barbed wire and concertina enclosing the U.S. military complex commonly referred to as the Green Zone. The desert sun was rising over the Land of the Two Rivers, turning the Euphrates River blood red, which seemed appropriate considering the state of affairs in the country. A few blocks almost due west from the zone's heavily guarded main gate sat a series of low sand-colored apartment buildings taking up the better part of three square blocks. In Los Angeles or New York, it would have been called "urban blight." The only information provided me was that our snitch claimed that the hostages were being held somewhere in one of the apartments.

Not much to go on. My crew and I were to prowl around and report any suspicious activity. I circled the complex, carefully panning for guards or for somebody who looked out of place, acted suspicious, or stayed in one location too long. Bodyguards almost always accompanied important insurgency leaders to protect them. Ironically, the presence of an armed goon hanging around a front door or on a porch or the rooftop of a house often tipped us off to the presence of a worthwhile target inside. Gunmen would most certainly be detailed to guard hostages if they were present.

I soon focused on a guy who stood in relaxed parade-rest in front of a door to an apartment in almost the dead center of the complex. While people came and went elsewhere, all astir in a constant bustle of morning activity, this man remained stiff and watchful at his post. He was a tall, beefy individual wearing a red-and-white-checkered *keffiyeh* and a *thawb* with enough billow to conceal a knee mortar or a small machine gun. There was something inside that room that he didn't want the world to know about.

I fed live video to the unit we were supporting, an army Brigade Team. I was asked to stand by and keep my eyes peeled for other people coming and going.

That assignment took up most of the day. The hot sun arced across the sky from east to west. Shadows lengthened. My guy sometimes swigged water from what appeared to be a goatskin. He smoked cigarettes. He had a snack about midday. He took a few steps to one side, lifted his *thawb*, and urinated against a wall, unaware that he was being watched all the while.

That was the extent of his activity. No one came or went. Finally, the unit we were supporting advised that they had decided to raid the apartment. Apparently, they had received additional information.

By that time, my relief had reported on-duty at Nellis. I yielded my seat to him since nothing hot was going on at the moment. I nonetheless hung around for another hour or so waiting for the operation to kick off. I had to know if our hostages were alive and inside the apartment. Plus, I longed to see Zarqawi nabbed or dispatched with extreme prejudice.

In Nevada, the sun was coming up. I had been awake all night. I yawned and at last opted to make a run home for a couple of hours' sleep since it appeared that nothing was likely to happen until after midnight, Baghdad time. The MCC promised to call me if the hostages were recovered.

Trish awoke me at noon. It had been on the news, she said.

"They got them!" I exclaimed, jumping out of bed to ring the POC.

Two events had unfolded while I slept. First, the good guys raided the apartment. It was a dry hole except for the bodyguard, apparently posted there to decoy us from elsewhere. That didn't make the news. What led CNN's early report was Jack Hensley's beheading. In a Web video, one of the masked killers, presumably Zarqawi, read a long, rambling statement.

"Thank God, the lions of the Tawid and Jihad have slaughtered the second American hostage at the expiration of the set deadline. . . . Oh, you Christian dog Bush, stop your arrogance. The Mujahidin will give America a taste of the degradation you have inflicted on the Iraqi people. The British hostage will face the same fate unless the British government does what is necessary to free him. . . ."

I stood at our back window for a long time staring out over the fenced-in piece of desert that served as our back yard, batting back tears of anger and frustration. Trish came up from behind and put her arms around my waist.

"I feel so helpless," I murmured. "I tried, Trish . . . I tried to help save them."

"You did what you could," she commiserated.

"It's never enough, honey. It's never enough. And what we're doing isn't working."

Her arms tightened around my waist.

CHAPTER 10

Red Bulldozers

VETERANS RETURNING FROM THE Vietnam War were stunned when they got off airplanes on U.S. soil and had to change out of uniform in the airports in order to escape angry protesters railing against "baby killers" and "warmongers." Overseas, they had been more or less insulated from what was going on back home. The same was true to a lesser extent of servicemen on the ground in Iraq or Afghanistan; they were too busy surviving and fighting the war to pay attention to criticism readily available through the Internet and phone calls home. *My* war, however, was a different matter.

I lived in two worlds, one at war, the other relatively untouched by it. I felt the barbs personally as the world began to learn about the use of unmanned warplanes. And there was plenty of criticism at the very concept of armed but remotely piloted aircraft. Columnist Nat Hentoff, of the Newspaper Enterprise Association, led the charge.

"My concern," he railed, "is that drones are being operated in a framework which may well violate international humanitarian law and international humans rights law. . . . I would like to know the legal basis upon which the United States is operating . . . who is running the program, what accountability mechanism is in place in relation to that."

Clearly Hentoff did not understand how carefully we utilized the Predator and the layers of responsibility that checked its misuse. It must not have occurred to him that while it may seem like science fiction, we treated RPA technology as an extension of airpower. We followed the same rules of engagement and used the same procedures as all other aircraft, manned or unmanned, that employed weapons in support of the fight on the ground. To us, the Predator is a longer-duration, lightly armed (and much less survivable) version of an F-16—with the benefits of persistence, global distribution of video and data, the ability to leverage the entire intelligence apparatus through ground communications links, and the ability to think clearly at zero knots and one G.

"Our credibility as a nation of law is being questioned around the world," Hentoff continued.

Philip Alston, the United Nations special rapporteur on extrajudicial executions, raised the issue on the floor of the UN.

In no way was it like Predator ran wild killing people at will. Unmanned aircraft were *not* a separate military power. Like special forces or navy SEALs, they were an extension of the general military and as such accountable to the laws of armed conflict within a military chain of command subservient to civilian oversight.

War is not sport. Critics missed the point that it was our duty to make war *unfair* and as much to our advantage as possible. From the perspective of a pilot who flew missions every day in direct support of troops on the ground in Iraq and Afghanistan, I knew that Predator *saved* lives, both American and civilian, by its ability to monitor situations so that operations could be deliberate instead of reactionary; by its capability to coordinate almost instant and well-informed expertise from the ground and around the world to make strategic and tactical decisions; and by a command philosophy that encouraged the reduction of collateral damage. Other methods of warfare could be, and often were, much more destructive.

The incident of the red bulldozer in Fallujah was a prime example.

Zarqawi and his cutthroats knew that Wyatt Earp would ride into Dodge sooner or later to tame the town. In anticipation of that day, they used bulldozers and other construction equipment to barricade off key avenues. Junk cars, piles of old tires, scrap steel, building materials, anything available was employed toward the goal of creating a maze of dead ends and cul-de-sacs. It was classic guerrilla strategy to bog down opponents with an obstacle course and then pick them off in the confusion. The so-called Skinnies in Mogadishu did it to our army Rangers. Ahmed Shah Masood, one of the most outstanding guerrilla fighters in history, blocked off mountain passes to trap advancing Soviets in Afghanistan, sometimes decimating entire columns. Zarqawi thought to do the same to marines.

Our fast movers blew up the makeshift barricades whenever they could safely do so without inflicting collateral damage. Citizens and Iraqi police in the city were afraid to do it themselves and draw the ire of the insurgents. Nonetheless, no matter how many barricades we bombed, Tawid and Jihad erected them faster than we could destroy them. Special Ops requested that I locate a red Caterpillar bulldozer being used in the insurgents' effort to shut down the city and plant a Hellfire in the driver's seat. Simple enough. How many places could you hide a red bulldozer?

My crew and I began hunting, sailing lazily overhead and scanning crowded streets for not only the dozer but also for signs of Kenneth Bigley, the kidnapped British contractor about whom we had heard nothing since Hensley's beheading. I soon came across the Cat lumbering down a side street in an attempt to maintain a low profile. Even though the poor bastards had to be aware of the reality that they were having an increasingly tough time hiding their activities, they apparently never considered our almost-invisible spy in the sky. Some of the men found beheaded or hanging from light posts were undoubtedly killed because

Zarqawi thought they were snitches spilling their guts in secret to the Americans.

The bulldozer's top speed was about ten miles per hour, which made tracking it a cinch compared to chasing some suicidal Iraqi in a Renault intent on becoming a martyr to collect his seventy-two virgins in Paradise. I often wondered what the virgins got out of the deal.

Rules of engagement (ROE) required that we endanger as few people as possible when we employed force, contrary to what Predator critics apparently believed. We followed the Cat through Fallujah for about a half-hour as it crawled along with no obvious indication of destination, waiting for it to stop somewhere in a clear area so I could nuke it without injuring bystanders.

I was so concentrated on the task that the sudden strident voice of a combat controller over the radio guard frequency almost caused me to jump out of my seat.

"*Marine Saber Two! Turn right heading two-one-zero. Traffic twelve o'clock at four miles. Copy, Saber Two? Turn right immediately.*"

Any pilot upon hearing a warning like that instantly began craning his neck. Saber Two was on a collision course in the air with somebody. As I wasn't *in* my aircraft, visibility was restricted to what I could see on my computer screen through live video—and all my cameras were aimed at the bulldozer and not at the surrounding airspace.

I raked my eyes across live video to the map screen, where I spotted a marine F-18 icon headed directly toward my Predator icon at about five hundred knots. I yelled at Senior Airman Cunningham, my sensor operator, "Pull off-target. Point the camera at the F-18."

At his speed, the marine pilot in the fast mover had mere seconds to react and avert a midair collision.

Had I been piloting a manned aircraft, I would have had several options—dive, slam the stick right or left in a hard turn, pick up speed, or slow down. Instead, my only recourse in the Predator was to switch off autopilot and initiate a slow, smooth bank of about thirty degrees off the F-18's heading. Any radical movement chanced losing the satellite link and therefore all control of the airplane. If that occurred, the Predator automatically would return to wings-level per its lost link profile and would remain in the F-18's path.

Airman Cunningham was already scanning the right-hand horizon with the infrared camera. I saw the F-18 in real time closing in on us at a phenomenal rate. He seemed to be turning away—but would it be soon enough?

I was so into the moment that every muscle in my body tensed for the impact. I leaned into the turn with adrenaline pumping. I couldn't have been more involved had I actually been inside the plane. I almost felt Instructor Moran over my shoulder. "*You're in that airplane, Captain Martin. Feel it!*"

For just a moment, I reverted to survival training instinct and thought about ejecting. What a hell of a wasps' nest it would be to parachute into downtown Fallujah, where wolves outnumbered sheep four to one.

Fallujah? If *I* ejected, I would find myself standing outside the GPS trailer in Nevada.

Closing at 830 feet per second, the F-18 almost stole my breath as it zoomed past within a few hundred feet. That was a close call at his speed. I sighed deeply in relief and slowly rolled the Predator back to wings-level. I shook my head and looked around in aftershock. Instead of being over Iraq fighting to save my ship, I was stationary again in my GCS at Nellis Air Force Base, 7,500 miles away. I glanced at Cunningham. All the blood seemed to have drained from his cheeks. He must have been right in there with me above Fallujah.

"Whew!" he exhaled. "I almost felt the wind when he went by."

We each managed a nervous laugh.

In the aftermath of our getting things back under control, double-checking our assigned airspace to make sure we were where we were supposed to be, and refocusing our cameras, the red bulldozer disappeared. We frantically searched up and down streets, back alleys, courtyards, and fields. Nowhere. It seemed to have vanished from the city. Cunningham and I went off-shift hours late without having located it again.

The next day I read a marine after-action report (AAR). An F-18—perhaps the same one that caused me to lose the Cat in the first place—had found the bulldozer near the edge of town building a barricade, along with a couple of dump trucks, a front-end loader, and about a score of men with shovels. None of the workers could be considered exactly innocent. Nonetheless, the marine F-18 pilot, in my opinion, went a bit overboard in his enthusiasm.

After receiving clearance on the target, he screamed out of the sky in a surprise attack, delivering four 500-pound bombs, two pods of 3.5-inch rockets, and, for good effect, wrapping it all up in five hundred rounds of machine-gun 20mm. The attack destroyed the red bulldozer and all the other equipment and left at least twenty people killed in action (KIA). That was what marines did. They broke things and killed people.

Irrational as it may have been, reading the AAR left guilt riding me like a cruel jockey. Although the mistake wasn't mine in the near-miss, I couldn't help feeling that, had I kept a keener eye on air traffic in my sector, the incident wouldn't have occurred and I wouldn't have lost the red bulldozer. One precise Hellfire could have taken out the Cat with little or no collateral damage. Certainly it wouldn't have wiped out almost a city block.

I scolded myself for being squeamish. There was, after all, a war going on. We had to be willing to kill the enemy and destroy his stuff if we expected to win. I had no problem with wasting some dirtbag who deserved it—but some of those men down there were likely poor people trying to earn wages the best they could in a city under siege. Their deaths were on my head and on my conscience.

By its very definition, war was brutal. In the long run, I consoled myself, Predator allowed us to be less brutal. Still, I needed to pay better attention.

***

The logic behind the U.S. mission in Iraq as I understood it, and as it was often explained, was rather simple and straightforward. At least it was in the "Squeeze Chart" briefings we received. Each of three concentric circles on the chart portrayed a segment of the population. The circle in the middle represented ordinary Iraqis who merely wanted a life and were mostly glad that Americans were in-country to help provide it. The next circle depicted the so-called moderates who, as in any country, drifted with the prevailing winds and could be convinced to move into the inner circle if the conditions were right. In the outer circle were the extreme fringes who beheaded captives, blew themselves up with Baghdad belts, planted IEDs, ambushed Americans, and terrorized those in the other two circles.

The aim of insurgents in Iraq was not primarily a military one; most of them knew they couldn't win with arms. At least not yet. Instead, they strived to apply pressure to the circle of moderates in order to swing them away from the political process. Those who had "rather fight than switch" often became corpses left in the streets.

The operational strategies developed by the Multi-National Corp-Iraq hoped by attrition to eliminate the lunatic fringes by hunting down leaders like Zarqawi and his lieutenants in order to relieve pressure on the moderate majority and allow it to move toward the circle in the middle, thereby permitting political progress. Providing stability for Iraq's first free election, coming up in less than three months, became a major objective for all coalition troops. We escalated actions throughout the war zone to ensure this stability.

Fast movers with JDAMs smashed the sites code-named *El Campo* and *Richardson*, those being the safe house and the restaurant from which had originated the Predator-car-helicopter chase across the desert outside Fallujah. We also targeted as many mortar and rocket teams and local cell leaders as we could whenever they got too careless. What we really wanted was the big fish, the leaders and organizers, the *committed*. Taking out the little fish was turning into an endless task. Any number of unemployed men were always willing to visit their local mosques, collect a mortar tube and the equivalent of about a hundred dollars U.S., and then go out and shoot at somebody. For every one or two of these we nailed, three or four others stepped forward to take their places.

Sometimes I found amusing the lengths to which the bad guys went in their attempts to avoid detection. They figured we located their safe houses and hiding places either through informants or by tailing them in other vehicles. It apparently still hadn't dawned on them that we were able to watch them from a couple of miles or so up in the sky.

One afternoon my crew and I picked up on a van as it left a recognized safe house in downtown Baghdad. It threaded its way along side streets to a main thoroughfare congested with traffic. Suddenly, it stopped. Three or four guys jumped out and darted into another car parked at the curb. The van, now occupied only by its driver, hooked a turn at the next corner to attract any tail that might be following. The other car—it would become apparent that it was full of jihadists—went in the opposite direction, weaving in and out of traffic. It seemed that vehicles in Iraq came equipped with only two controls—gas pedal and horn. And since there were few traffic signal lights or stop signs, everyone drove like a kamikaze pilot high on crack.

Of course, I kept on the car. The van was just a decoy. My guys changed vehicles two or three times more before they reached their destination—another safe house we code-named *Ma Bell*. A couple of other cars were already there. The passengers got out and looked around. One stepped onto the street and looked both ways. Then they all went inside, satisfied with their resourcefulness.

I was rather impressed. They were clever. But not clever enough.

I radioed in, fed the video and information to the ground commander, and then waited. About ten minutes later, Iraqi police and some of our army troopers showed up like gangbusters. They rounded up the surprised conspirators and hauled them away. I doubted they ever figured out how we found them so quickly.

"Goofy bunch," I snickered.

How much more progress we could have made had we a dozen Predators orbiting to pinpoint raids like that every day. That would come later.

I remained a voyeur in the sky snooping on peoples' lives. Increasingly cynical, I was suspicious of everyone.

Night had fallen on Iraq an hour or so after my shift began when I noticed a group of men sitting around on their haunches in a circle at a park in Sadr City. What first attracted my attention was the glow of cigarettes. Virtually everyone in Iraq smoked from the time he was ten years old, including women, although they did it secretly. The burning cigarettes became spectacular balls of radiation when viewed through an infrared camera, like hands on fire. I didn't know if the gathering was a party, a terrorist cell meeting, or just a bunch of men smoking and joking. We decided to keep an eye on them for a while.

I received a crash course in Iraqi cultural behavior over the next hour. Every once in a while, one or two of the men suddenly jumped up and danced around inside the circle. Several held hands, while others nestled in each other's arms like lovers for warmth against the cool night air. I knew enough about Muslims by now to realize they weren't gay; Muslim fundamentalists beheaded gays. Men in Arab countries were simply affectionate toward one another. I had often seen men sleeping together on the flat roofs of houses. I assumed it was purely platonic. Part of the culture.

One of the men eventually got up off the ground and walked over to a nearby shack. I thought I finally had them. He was going for weapons.

He returned with folding chairs. He went back and got some more, enough for everyone. They smoked and joked some more. I kept hoping somebody would pull out a rocket launcher. At least it would mean I was making good use of Predator's time and resources. Besides, blowing up things was much more interesting than watching men sit around in the dark smoking cigarettes, dancing, and holding hands.

CHAPTER 11

Elections Afghani Style

ALTHOUGH TERRORISM HAD RARELY been successful at advancing political goals or building a new society, it was making the pacification process difficult in Iraq and Afghanistan through intimidation of coalition partners. For months, al Qaeda had been bragging on its websites how it broke Spain's backbone with the train bombing in Madrid, thereby persuading Spanish voters to change governments and withdraw Spanish troops from Iraq. Since April 2004, a period of little more than six months, terrorists had kidnapped some two hundred civilian hostages to back up their various demands and force coalition governments on the defensive. While the tactic worked against some countries such as Spain, citizens who were seized from nations that refused to negotiate with terrorism did not fare so well.

There were protests and antiwar marches in England after Zarqawi finally beheaded the last of the three contractors, Briton Kenneth Bigley. For weeks we had continued to search all of Fallujah and Baghdad for him, failing to come up with a single lead or clue as to his whereabouts until it was too late.

That was another time when, frustrated and angry, I stared out the back window of our little house in Las Vegas. Rather than the failure weakening my personal resolve, however, it only deepened my growing contempt for Zarqawi and his ilk.

Meanwhile in Afghanistan in early October 2004, Afghans thronged to the polls to vote for the first time in Afghanistan's long history. Out of a slate of twenty-three candidates, Hamid Karzai was elected for the presidency by commanding 55 percent of the nation's votes. Elections for Parliament were still pending.

Iraq was rapidly approaching its own first free elections in January. Baghdad as well as Kabul was aswarm with diplomats and other foreign election officials. Troops in both countries were warned to be alert for increased violence throughout November and December in Iraq, and through the rest of the year and into 2005 in Afghanistan.

As we had only one Predator flying in Afghanistan at the time, there was no way we could conduct surveillance over every potential trouble spot. All through October, even past Karzai's election, crews out of Nellis soared shifts above Kabul, constantly panning the capital. Several people were shot on election day, presumably by Taliban attempting to thwart the procedure, and a bomb exploded downtown. Many fearful and cynical voters, especially women, simply stayed home. While opposed to all elections, the Taliban particularly resisted women voting. Let 'em vote, next thing they'd be wanting to work and drive cars.

Sporadic violence and intimidation persisted even after Karzai's election as jihadists attempted to prevent his being inaugurated. Homegrown insurgents and al Qaeda cells were obviously paying attention to the press that Zarqawi received as a result of his kidnapping of foreigners. Encouraged, they decided to take a shot at it themselves. No Westerners had been abducted in Afghanistan since the fall of the Taliban in 2001.

On the morning of October 28, 2004, three officials of the joint UN-Afghan electoral body left their offices in the capital to return to the upscale Wazir Akbar Khan neighborhood, where many aid workers and diplomats lived. On a dirty side street, a dark-colored four-wheel-drive vehicle occupied by five men in military uniforms forced over the white SUV clearly marked as an election vehicle. "Soldiers" dragged the two men and one woman from the SUV and trundled them into their own car. United Nations security staff found the election workers' vehicle abandoned with its doors locked later that same afternoon.

The victims were Annetta Flanigan, from Northern Ireland; Angelito Nayan, from the Philippines; and Shqipe Hebibi, from Kosovo. When last seen, abductors and captives were heading in the direction of Paghman, a district in western Kabul Province considered rife with bandits and cutthroats. Intel thought they might go into hiding there in one of the small villages.

Not unpredictably, again aping Zarqawi, the kidnappers put the victims, frightened and cowering, on the Web. I expected their next appearance to be a big finale in which they parted with their heads—unless we rescued them first. An organization known as Jaish-al Muslimeen—Army of Muslim—was already threatening to execute the trio unless the United States and the Afghan government released twenty-six militant prisoners and withdrew all U.S. forces from the country.

"There are some of our members who have hard-line views on the issue, but there are others who have moderate views," equivocated a man who identified himself via telephone as Mohammad Akbar Agha, leader of the Army of Muslim, a splinter of both al Qaeda and the Taliban. "The hardliners say we should get rid of the hostages," the man said. "The others say we have the ability to keep the hostages for two years."

We Predator crews concentrated our search in western Kabul Province during the days immediately following the seizure. It was mostly a rural land dotted with tiny farm villages. We stood a better chance of finding the election workers in this kind of terrain than we had had in locating Armstrong, Hensley, and Bigley in the teeming cities of Fallujah and Baghdad. But then, on second thought, Osama bin Laden was supposedly hiding out somewhere in the mountains of Afghanistan—and we hadn't found him yet.

At three o'clock in the morning Afghan time the day after the abduction, I flew my airplane westbound from the Pakistani border to investigate a village on the outskirts of Kabul. A recent snowfall mantled the ground in white, which reflected the soft light of a three-Bobby moon. It was one of those cold, dark nights that drove sane people home and to bed. Although it was only shortly after noon in Nevada, I got the yawns just looking at all that snow and darkness.

Not everyone in the village, I soon discovered, was home in bed with visions of sugarplums dancing through his head. My infrared picked up the movement of four men digging a van out of snow behind a shack. I could almost see their breath vapor. Ice crystals glistened under the moon as they scraped the windshield clear. Then they clambered into the van and after two or three tries managed to get out of the driveway and chug off down the road. Something must be mighty important to draw them out on a night fit for neither man nor Santa Claus. My first thought was that they might have something to do with the kidnapped election workers.

I followed. Their headlights splashed cones of light against the night. After a mile or so, the van turned off the road toward what appeared to be a warehouse. A fellow manning the guard shack opened the gate. Workers were waiting to load the van with boxes, some of which they strapped to the roof when the cargo bay was full.

Once loaded, the van with its occupants returned to town, where it stopped at various houses to deliver the boxes. Maybe it was ammo-delivery day, sort of the way milkmen used to drop off bottles of milk on doorsteps in Norman Rockwell's America.

No doubt I had something going on, but I was beginning to doubt it had anything to do with the abducted workers—until I saw two people exit a house and start toward the van. It appeared that one person was leading the other against his will. Of course, ice and snow made the walking slippery and the footing unsteady, which meant that the one might merely be assisting the other. I changed my mind again, however, when men jumped out of the van and pushed the reluctant subject into the back seat.

Wearing snow chains on its tires, the van roared out of town on the icy road, climbing into the mountains by the time the sun began to spread the snow with a pink morning tint. I used the chatroom to notify ground commanders of developments while feeding them live video.

Keep on it came the chatroom response.

We followed the van along precipitous mountain passes for the rest of my shift. Into snowy ramparts where Osama bin Laden might still be hiding with the Abominable Snowman. I was still on its tail when my relief pilot, Maj. "Scooby" Wachoski, entered the GCS to take over. I reluctantly turned over control of the aircraft to him and his sensor. The "chase" had already been going on for hours and looked to go on for several more. I could always find out tomorrow what happened.

"We think one of the kidnapped election officials might be in the van," I briefed Major Wachoski. "We still have about ten hours' fuel left. Air assets are on standby for a rescue attempt if it turns out that way."

"Don't worry, we won't lose him," Scooby promised, adding, "You still have two hours left to vote before the polls close."

It was presidential election day in the United States, November 6, 2004. From the snowy mountains of the land once conquered by Alexander the Great and Genghis Khan, I stepped out into late sunshine in Nevada and a temperature that still hovered around ninety degrees. It occurred to me that nobody in America would have to worry about getting shot, blown up, or kidnapped on the way to the voting booth.

In servicing two war fronts with a limited number of Predators and crews, we were necessarily drawn back and forth between the hunt for the kidnapped election workers in Afghanistan and the growing anticipation in Iraq of another push into Fallujah. The van I followed through the snow into the mountains turned out to be unconnected to the abductions, although the occupants *were* Taliban insurgents. I never found out what happened to them, I was so busy elsewhere. That was the nature of this strange war fought in the air by remote: a lot of strings were left hanging.

For twenty-seven days, Predator flew twenty-four-hour shifts over Kabul and the surrounding countryside trying to locate where Annetta Flanigan, Angelito Nayan, and Shqipe Hebibi were being hidden. Clouds developed and produced snow some days and we were grounded. Otherwise, day after tedious day we kept after it.

Lawless Afghanistan remained in the grip of warlord Taliban militia and al Qaeda in spite of Karzai's election to the presidency and the presence of American and NATO soldiers. More than a thousand people had died from political violence so far this year, including at least a dozen election workers. A few days after the seizure of the three poll officials, a suicide bomber armed with grenades blew himself up on a busy Kabul shopping street, taking an American woman and an Afghan teenager with him and wounding three NATO soldiers. I figured that if we didn't find the kidnapped victims soon, they'd end up like Zarqawi's hostages in Iraq. From what I could find out, Army of Muslim leader Mohammad Akbar Agha was cut from the same bolt of cloth as Zarqawi. Just as ideological, although perhaps not as ruthless. Zarqawi would never have waited twenty-seven days before inviting the world to one of his little beheading parties.

Behind the scenes, diplomats from around the globe kept up negotiations with the kidnappers while at the same time the UN insisted it did not bargain with terrorists. One afternoon, I reported to the POC for duty and discovered an electric air of excitement. People were laughing and slapping one another on the back. Major Scooby Wachoski wore a huge grin when I relieved him at his cockpit seat. He jumped up and pounded me on both shoulders.

"We got 'em, Matt."

He caught me by surprise. "Got who?"

"The election workers. All three of them are safe and sound."

I never expected it to end that way.

"Did *we* do it?" I asked.

By *we*, I meant Predator. I was totally committed to the use of unmanned warplanes, having experienced Predator's capabilities in reconnaissance as well as being a force multiplier and a means to achieving ends less violent than carpet-bombing. Each Predator success purchased the trust of more and more military skeptics.

However, neither Predator nor our rescue forces were involved in the recovery. The UN apparently received a phone call saying that Flanigan, Nayan, and Hebibi had been released at a park in Kabul. Sure enough, there they were, all of them looking healthy, happy, and fairly normal considering the four-week ordeal they had endured.

"Their kidnappers must have had a change of heart," Scooby said.

Terrorists didn't *have* hearts. The story came out piecemeal over the next few days, even though UN representatives ordered the freed hostages to say nothing about how and why they were released. Afghan Interior minister Ali Ahmad Jalali denied that any deal had been cut with the kidnappers to pay either ransom money or let Taliban prisoners go.

"No prisoners were released, no money was paid, no demand was made of the hostage takers," he insisted. "And to my knowledge, no other parties paid money."

A UN senior official backed him up: "We are still trying to figure out what happened and why they were freed."

I had my doubts that they were telling the truth. It would be a huge embarrassment if the UN connived to pay off the kidnappers when its official stance was not to negotiate with terrorists. Jihadists rarely seized hostages and then released them without receiving something in exchange.

The mechanics of how the three escaped their fate was ignored for the next several days in the afterglow of their being alive and well. It seemed the entire world celebrated their release. Global headlines declared Annetta Flanigan a heroine.

Their first night in captivity, the hostages thought they were going to be shot when they were taken to a shed on the outskirts of Kabul and forced to lie face down on the dirt floor. Although frightened and chained in the freezing cold of her makeshift prison, Annetta used charm to persuade their kidnappers to spare their lives. Smiling and refusing to panic or lose her temper, she kept talking softly to the gunmen, asking about their families and telling stories about her own home. Smart cookie. She realized it was more difficult to murder someone you liked and came to know as a human being.

For the first two weeks, the hostages were rotated between four separate houses in the capital. Eventually, they ended up in an abandoned building on the outskirts that offered little protection from winter winds. However, their captors began to treat them better. They were permitted to brush their teeth, play cards, and read books. They were also given clean underwear and allowed outside in the fresh air to sit with their guards and watch children flying kites. Improved treatment seemed to indicate that negotiations might be going well.

In fact, as it turned out, the UN *may* have been negotiating after all. A British hotelier named Peter Jouvenal, who owned a guesthouse in Kabul, revealed that he served as a go-between. "I was involved in the negotiated release of the three UN officials," he admitted.

According to Jouvenal, he traveled to Pakistan and the city of Peshawar to meet with Mohammad Akbar Agha. Agha himself told news agencies that the UN and the government of Afghanistan had submitted to his demands to release twenty-four Taliban prisoners. Money also changed hands—$1.5 million.

Subsequent intrigue produced a curious and amusing side note. It seemed that the Army of Muslim broke up in a row over the ransom money. Mohammad Akbar Agha, not as dedicated to the cause as we supposed, absconded with the cash and was last seen on his way to Dubai.

"It's going to be bloody when the marines return to Fallujah," I confided in Trish one night. "There are going to be a lot of our kids killed in that damnable city. I have to do everything I can to minimize our casualties."

Trish observed how I looked haggard, worn, how I wasn't sleeping well and not laughing so much anymore. Perhaps it was difficult for someone not involved with unmanned warplanes to comprehend how pilots and sensor operators 7,500 miles away from the action could become so involved, so personally invested in the war. After all, we weren't getting shot at, wounded, or killed. Even if we got shot down, we didn't *really* get shot down.

What they also failed to understand was that I *knew* people down there. Each day through my cameras I snooped around and came to recognize the faces and figures of our soldiers and marines, unbeknownst to most of them. I sometimes chuckled over their youthful pranks and hijinks when they were off-duty and in secure areas. I cried with them as well whenever they lost a comrade and they huddled together with their arms over one another's shoulders. Looking down on all this, seeing the foibles and courage and decency, and all the various behavior, emotions, and ways of mankind at its best and at its worst, I truly felt a bit like an omnipotent god with a god's seat above it all.

"A lot of lives would be saved," I told Trish, "if we could find Zarqawi."

The day of the town tamer was rapidly approaching. Helicopters dropped leaflets into Fallujah warning people in the center of the "Squeeze Chart" to evacuate. Marines tightened their seal on the city until it was as tight as a condom on a large cucumber. None of the insurgents was going to escape this time.

As November began to unfold, streams of refugees and displaced residents fled in vehicles and donkey carts, on foot and pushing wheelbarrows, clutching children by the hand and looking back with tears on saddened faces. A scene that had played out countless times throughout the history of nations and people. I empathized, as I could imagine how it might feel were I commanded to evacuate Las Vegas through a ring of foreigners preparing to go in and raze the city in order to get rid of thugs and street gangs.

Probably 80 to 90 percent of Fallujah's population fled ahead of the pending operation. Those who remained behind would be considered hostile and subject to action against them. About five hundred "hardcore" and perhaps two thousand "part-timers" fought marines during Operation Vigilant Resolve, the first battle of Fallujah, which began in April 2004 with the murder and mutilation of the four Blackwater security specialists. During the seven months since then, while marines cooled their heels on the perimeter and gave the Iraqis a chance to tame their own city, insurgent strength and control of the city had grown to such an extent that a "senior U.S. official" told *ABC News* that apprehending or killing Abu Musab al-Zarqawi was now of "the highest priority." About five thousand men, mostly Zarqawi's non-Iraqis, were prepared to fight to the death during the Second Battle of Fallujah.

Six battalions of army, marines, and IA forces—over ten thousand warriors—were coiled and ready to strike under the cover of darkness. Navy Seabees shut off electrical power at a substation northeast of the city. Air force and marine fighter-bombers began to pound key objectives to soften insurgent strongholds. From high in the air, I watched the flash and flicker of explosions. Palls of boiling smoke gradually pulled a curtain of haze over the city.

No more Mister Nice Guy.

Omnipresent over Fallujah, witness to it all, Predator pilots and crews were assigned to pinpoint targets for the fast movers. We sought out elevated sniper positions and fortifications on the streets and on roofs, groups of gunmen, and "technicals" with their mounted machine guns in the beds of old Toyotas. One afternoon I observed a band of insurgents brazenly conducting live-fire exercises in a central city park. These fools never seemed to learn that we could see them. I called on an F-18 to pulverize the park with bombs and machine guns and send the bad guys to Paradise to collect their virgins.

I knew that time was short when Col. "Race" Bannon, the mission commander at Nellis, showed up for the preflight briefing on November 7. He carefully walked pilots and sensors through air traffic control procedures that would go into effect over Fallujah once the fight started. A lot of aircraft would be using a relatively small amount of airspace, creating a high-risk scenario.

After all the weeks of run-up leading to the battle, I anticipated finally getting in there and getting it over with. I knew that soldiers and marines on the ground felt the same way. Tension before an action often proved more stressful than the actual thing.

"Tomorrow," Colonel Bannon concluded, "is d-day. Get ready."

CHAPTER 12

Operation Phantom Fury

OPERATION PHANTOM FURY AGAINST Fallujah kicked off under cover of darkness in the early hours of November 8. Thunderstorms crashed and banged over the unlighted city, merging lightning strikes with the muzzle flicker of tank guns, the flash-bang of explosions, and streams of tracers from machine guns. It was mind boggling how hard rain could fall over terrain regarded as desert most other months of the year.

Unlike other predators such as lions or wolves who hunted no matter what, Predator operations were limited by inclement weather. The battle started without us, our small fleet of unmanned warplanes remaining grounded at Balad Air Base outside Baghdad due to low-wing loading, symmetrical laminar-flow wings that were prone to ice up upon exposure to cold moisture, lack of deicing equipment, vulnerability to strong winds, and the necessity to keep the gyro-stabilized satellite dishes on the planes properly aligned. Restless and frustrated at being relegated to the sidelines of a major fight that had been brewing for months, pilots, sensors, and other personnel associated with Predator operations gathered on the POC floor to monitor chatroom discussions or garner second hand whatever news we could from Fox or CNN. Zarqawi seemed to be on everyone's mind.

I overheard someone mutter with raw passion, "We've got to kill that sonofabitch *this time*."

A certain urgency accompanied the launching of the offensive because Iraq's first free election was rapidly approaching. Each day that the insurgency kept its stranglehold on Fallujah made it that much harder to get the democratic process to take. Marines would certainly have preferred to fight in good weather, but they couldn't afford to scrub the operation after so much planning and wait for a sunny day. Fallujah *had* to be pacified—and pacified quickly.

The battle launched in high gear and then built up momentum. It began with troops attacking and securing the main train station in order to use it as a staging point for follow-up forces. By early afternoon, marines were fighting their way into the Hay Naib al-Dubat and al-Naziza districts, followed by navy Seabees, who bulldozed streets clear of insurgent roadblocks and debris left by artillery and air bombardments. CNN was already proclaiming how the offensive was producing some of the heaviest urban combat since the Battle of Hue in Vietnam during the 1968 Tet Offensive.

For the last several weeks, U.S. F-18s, helicopter gunships, and Predators had repeatedly struck insurgency nests and battlements in the buildup to the operation, generating a constant stream of propaganda from the main hospital in Fallujah about how Americans were killing and maiming hundreds of civilians. Kids who had fallen off bicycles and old men with tuberculosis coughs were wrapped in bloody bandages and paraded before cameras as proof of American atrocities. Al-Jazeera and the European news media were all too willing to pass on reports that were detrimental to the United States. Even our own news outlets in the United States went with the inflated collateral casualty numbers; most impartial journalists didn't know better because they were holed up in Baghdad until after the danger passed.

Controversy raged around the globe when one of the first objectives our troops secured was the hospital. Nonetheless, seizing the hospital allowed independent journalists to verify that casualties among women and children were far fewer than our enemies claimed.

Iraqi Sunnis had dominated government, business, and most civic institutions under Saddam Hussein, although they comprised only 25 percent of the population. Shiites and Kurds made up the other 75 percent. Having little interest in establishing a government that was bound to marginalize them, Sunnis had kept the insurgency fed with recruits, weapons, and materiel. From my detached viewpoint grounded in Nevada, I watched the Sunni reaction to Phantom Fury with dismay if not exactly with surprise.

Almost as soon as the assault in Fallujah began, Sunni political parties announced they were withdrawing from participation in the interim government. At the same time, the Muslim Scholars Association of Sunni Clerics, which claimed to represent three thousand mosques in-country, called for a boycott on the national elections coming up in January. These actions spawned a sudden sympathy escalation of IED attacks and ambushes in Baghdad that threatened to spill even more violence into the country's major cities. Things were turning into a real can of worms. It seemed that as soon as we stomped out one fire here, another blazed up over there.

I telephoned Trish from the POC and begged off dinner, telling her I would grab a sandwich later. None of the other pilots and sensors left either. We anxiously hung around awaiting news on how the operation was going, not departing until hours after our shifts ended and weariness and crew rest rules forced us to go home to catch a few hours' sleep.

My shift had begun in the late afternoon in Nevada, predawn in Iraq. I had been up all night. I drove home in sunshine. It was getting dark in Iraq and the weather was starting to break. By the time I returned to the POC for my next shift, we should be cleared once more to fly our nation's latest and greatest weapon back into the Global War on Terror.

Fighting a precise and low-collateral-damage war was much more difficult, stressful, and dangerous than going in and leveling the place until you were the last man standing, which, historically, was the way it had always been done. At no other time in history had a military force attempted to fight a war while at the same time going as far as the United States had in avoiding collateral damage. We *evacuated* Fallujah prior to initiating combat, foregoing the military-essential element of surprise and giving the enemy time to dig in, arm, and prepare himself for an optimum defense. Would that have happened in World War II, Korea, or Vietnam? Would the French have done that in Algeria?

Our politicians and higher military commanders seemed particularly risk averse and overcautious. We seemed to have concluded that the way to change the hearts and minds of the Iraqi people was to keep killing the enemy, but to try to be humane as we did it. At that point we had not yet shifted to a classic counterinsurgency strategy where building up the society was the main operational approach and hunting down the enemy was secondary. We knew that we had to "clear." But we hadn't yet learned how to "hold" and "build."

I certainly understood the policy of disrupting civilians as little as possible and harming as few as necessary. I would never forget the elderly man who walked by the brick wall in Sadr City just before I nuked Rocket Man. However, there came a point when we could be too cautious, too careful, too timorous.

That point seemed to have been reached on the second day of the battle when the thunderstorms moved on and we unleashed Predator. Only fighting-age men remained in the city. The elderly, the infirm, women, and children—those not prone to want to fight Americans—had moved out. I had no idea where they all went. Perhaps to the homes of relatives, to Baghdad or Nasiriyah, to refugee camps set up in tents underneath palm trees. A twenty-four-hour curfew in Fallujah was supposed to keep the few who ignored the evacuation orders confined to their homes. Anyone discovered in the open was presumed to be a bad guy and therefore a prospective target.

It was five o'clock in the morning in Iraq and an hour or so before full daylight when I began my shift by providing force protection for a column of armored infantry as it proceeded house to house hunting insurgents and kicking in doors. The line of Abrams tanks, Bradley fighting vehicles, and Humvees stretched down and around several city blocks like a giant mottled serpent with a poisonous bite. The action was slow. Marines were finding tunnels dug underneath the city, mostly between mosques and schools, which allowed fighters to scurry about like rats from place to place without being seen. Apparently, they were using these tunnels to avoid confrontation with heavy forces such as armored infantry and thereby focus their attention on plain infantry or supply trains.

When the column commander no longer required our services, we moved on to the next assignment—a report of hidden enemy artillery pieces along a riverbank. Using infrared in the predawn, we soon spotted four men walking along a street in flagrant violation of the curfew. None was armed, but I knew they had to be up to something. What most aroused my suspicion was the way that three of the four plodded along with their hands behind their backs while the fourth in the rear walked in typical Arab fashion with arms swinging.

We soon concluded that the three were prisoners being herded along by a guard. I watched for more than an hour while they wandered from house to house, as though seeking someone or something. This was still an unsecured area, which meant there were no friendlies nearby to check them out.

Finally, they entered a residence and didn't come out again, either settling in out of sight for the coming daylight or for something worse. Like an execution. There was little I could accomplish by watching an inactive house. I instructed my MCC to report the location so it could be code-named and investigated as soon as troops moved into the sector. Then I moved on.

The sun rose in a clear sky. There were no remaining signs of yesterday's thunderstorms. The sandy soil sucked up rain almost as soon as it fell. Heat and aridity did the rest.

On normal mornings, people thronged city streets going to work, to school, to mosque, and visiting neighbors, as well as preparing IEDs, shooting at Americans, and assassinating one another. Today, Fallujah looked as desolate as a scene from *The Day the Earth Stood Still*. Cows and donkeys and dogs and poultry roamed the city freely, although street after street had no pedestrians, as though no humans lived here.

Gradually, Fallujah began to awaken from its slumber. A machine gun hosing down a position, a firefight there, some explosions in a cul-de-sac making ugly greenish black clouds, marines scrambling into an alley.

Once more searching for artillery and antiaircraft along the river, I came across a grove of date palms that had been splintered, gouged, and cratered by our AC-130 gunships, mortars, and 105mm artillery. Nearby buildings had all been bombed down to standing walls only. Down through palm fronds I glimpsed the fractured landscape littered with fighting holes full of trash and perhaps corpses. It looked as though marines had been through last night to engage in close-up fighting.

I patrolled along the river for several more hours, encountering little that couldn't be handled by troops on the ground. I had expected much more action than this. I was about to give up and return to base at the end of my scheduled mission when we happened across some scruffy insurgent types loading weapons and ammo into the back of a pickup truck that was already hard-mounted with a machine gun. I tagged along when the pickup left until imagery analysts at one of the intel hubs confirmed that what my sensor and I saw really *was* contraband. Having been so conditioned to err on the side of prudence, they often took so long to make up their minds that the opportunity to do anything passed while they were still discussing it.

The pickup scooted to a second house nearby where other men were loading another truck with mortar tubes, RPGs, AK-47s, and boxes of ammunition. The analysts still vacillated, even though I could see the cargo. I wondered why anti-Predator pundits like Nat Hentoff didn't probe deeper and find out how careful we really were while using RPAs.

I burned up my chatroom screen back and forth with the imagery analysts.

All you have to do is look, for Pete's sake. That's a machine gun, that's a mortar tube. . . . What do you think they're going to do with them? Have a bake sale? They're going to use them against our guys unless we do something to stop them.

Axel Two-Seven, you're advised to stand by. Continue surveillance.

I asked my sensor, Senior Airman Netherly, if he wanted to stay in orbit with me past the end of our shift.

"Hell, yes."

The mission commander granted us a few more minutes. We had plenty of fuel and we were itching for a shot—if the analysts ever made up their minds about the obvious.

Not that I failed to see their side of things as well as mine. They would be held as accountable as Netherly and I should they make a bad call and we blew up innocents—even though all the innocents were supposed to be gone from the town. The ground commander actually had final responsibility to make the call; he was getting bad advice from hesitant support personnel. It seemed that no one wanted to make the final decision. I couldn't do it myself.

My replacement aircraft took off from Balad with a crew from another GCS and soon arrived on-station with me. We chatted back and forth while the insurgents below finished their business and split up, each truck speeding away in a different direction toward "the front." Netherly and I took one of the trucks, the other crew the second. I went down the Hellfire prelaunch checklist and called for a nine-line in order to spin up our Hellfires for maximum readiness. A nine-line contained the location of the target, the location of friendly forces, and the intent of the ground commander. I read the information back for confirmation.

My truck stopped in the street a couple of blocks over from where marines were moving stealthily from house to house. Jihadists jumped off the back of the truck and set up a mortar tube right in the middle of the deserted street.

Can you see it now? I taunted through the chatroom.

Stand by, Axel.

I had had enough hedging. Those marines two blocks over didn't have much time before shells started landing in their ranks. I skipped normal channels and took my case directly to the JTAC attached to the nearby marine unit. The ground commander was watching the video and ready to take action. However, he preferred marine artillery over air force Hellfires. He requested Netherly and me to call the shots by relaying target coordinates through the chatroom to a nearby 105mm Howitzer battery. I listened to the chatter on the radio as the battery opened for business.

"Fire mission. Enemy troops in the open."

"Enemy troops in the open."

"HE. Will adjust."

"Round out."

Something went haywire, I had no idea what. Maybe the insurgents had some kind of sixth sense. It was almost like they heard the 105mm round leaving the gun. Suddenly, for no apparent other reason, they abandoned the mortar tube where it was and scrambled back into the pickup. It was already burning rubber.

I waited for *splash*. Nothing happened. Again, I didn't understand. Somehow the marine fire direction controller (FDC) had made a mistake. No telling where the round actually landed, who it might have killed. I scanned across the city, looking for it. There were so many explosions I couldn't tell one from the other.

Axel, return to base.

The battle went on. I turned the aircraft homeward to Balad, checked off-station, and relinquished control of the aircraft to the on-site LRE to land and refuel. I left the GCS trailer dejected and discouraged. I felt we had failed today on both a human and a technical level. First of all, we neglected to kill bad guys with Hellfire missiles because we were afraid of making a mistake. Second, because we failed at our job, marine artillery may have wiped out people we didn't intend to kill. Maybe even our own troops. That, and two trucks full of insurgents escaped to use their weapons against our guys somewhere else in Fallujah.

CHAPTER 13

Leaving Nellis

MOST OF FALLUJAH WAS under control, pacified, by the last week of November—although fighting would continue sporadically until well after the Iraqi elections in January. Powers that be changed the name of the operation from Phantom Fury to Operation Al Farj, Arabic for "the Dawn." They wanted the Iraqis to feel like it was *their* venture, even though over half of their IA soldiers had deserted on the first day and the Iraqis had little part in planning or commanding and only a minor role in the fighting. Last count had 38 U.S. troops killed in action. Another 275 were wounded or injured. That meant limbs missing, vegetable-state head injuries, bodies maimed . . .

Whoever counted corpses tallied 1,200 enemy combatants slain. That meant we sacrificed one American life for every 31.6 insurgent lives we took. We were winning if you wanted to use body count as the measure.

Abu Musab al-Zarqawi was not among the 31.6. Intelligence people who interrogated prisoners learned that he was still alive and well—somewhere outside Fallujah. Everything indicated he had split for safer parts on the second day of the fight when it became clear that his bands were losing and that he personally risked death or capture. Apparently, it was okay for Zarqawi and others like him to recruit martyrs for the cause, but becoming one was a different matter.

From the air I looked down upon the scorched and scarred ruins of what, only weeks ago, had been a modern, bustling city. Fallujah was a wreck, every third building either destroyed or damaged. Broken windows, glass sparking sunlight in the streets, mud walls crumbled to dust, roofs collapsed so that you could see the contents inside, bullet-pocked walls, doors wrenched and blown from their hinges. I could have been looking at Berlin near the end of World War II. Iraq had not seen such destruction in any other of its cities.

Isolated pockets of enemy holed up in suicidal positions continued to resist—but everybody knew it was over. That was made even plainer when teams of U.S. Army lawyers began arriving in the city with suitcases full of cash to award to people with damage claims. In effect, they were insurance claims adjusters. Modern warfare could be totally insane. First, we busted up things. Then we ponied up for damages. Most of the people who collected were probably righteous bystanders. Others who showed up to collect cash were undoubtedly the same people we were trying to kill.

The war and what I saw and experienced of it was still too raw and personal to discuss with Trish, other than in general terms. She was my anchor with her soft red hair, patience, and understanding. Sanity in the midst of madness. I sometimes felt guilty about having her, being able to escape the war to go home to her when other soldiers, marines, airmen, and sailors were stuck in bunkers or sand holes in Godforsaken places halfway around the world. Although I was a participant much more so than at the beginning while flying out of Crete with the 193rd Expeditionary Reconnaissance Squadron, I was still as much a spectator as a participant.

That, however, was going to change. I was about to get even closer to the war.

"My orders to go to Iraq should be coming down shortly," I casually informed Trish one evening.

She was setting the table for dinner. Her only initial reaction was a brief freeze-up with her back to me so I couldn't see her face. She always strived not to be one of those service wives who went to pieces when her husband faced transfer to a combat zone. After all, she knew my profession when she married me.

"When?" she asked after she regained composure, trying to sound as casual as I.

"After Christmas. Probably January."

She went into the kitchen for something, her face still averted. She returned with a bowl in her hands, a smudge of flour on her nose. She stood in the doorway facing me, her eyes dry but her heart clearly unsettled. A spouse about to be left behind while her mate went off to war.

"Matt, you'll take care of yourself?"

After dinner, we curled up together on the sofa as we often did to watch TV. I had been looking forward to the documentary *Fog of War: Eleven Lessons from the Life of Robert McNamara.* I earned a postgraduate degree back when I had a lot of time to spare at the missile launch control center in the Middle of Nowhere, Wyoming. I wrote my master's thesis on the history of the arms race. The story of Robert McNamara was the story of the Cold War.

McNamara had been secretary of defense from 1961 to 1968 under Presidents John F. Kennedy and Lyndon B. Johnson. The war in Vietnam, McNamara said, was not centered on military conflict. Instead, he said, the true battle was being waged over the hearts and minds of the Vietnamese people. Establishing security in the country would provide time and opportunity for the democratic process to take hold—and the Vietnamese would surely be better off for it.

Years later, after the Vietnam War ended, McNamara rationalized that the reason we became so entangled in Vietnam was because we Americans believed our optimistic assessments of the strategic situation, even though they frequently had nothing to do with reality. American leaders viewed the war as an extension of the Cold War and the North Vietnamese as an extension of the Soviets. The North Vietnamese made a similar mistake in seeing the war as a struggle for freedom against a colonial/imperial power—an extension of the French. Each side was incapable of understanding the other—and the war was destined to rage on and on.

Although the Cold War was long over, what I saw in Iraq, and to a lesser extent in Afghanistan, seemed to be following the same pattern of optimistic assessments and misunderstandings. Was this then another war destined to rage on and on?

I felt myself growing even more cynical.

As we placed additional Predators into service in Afghanistan, we began to add to the national deficit due to hazardous weather flying conditions. Our sister squadron, the 17th Reconnaissance Squadron, lost one of its planes in the mountains. That same afternoon, the private company that handled LRE for us at a classified location crashed one of ours on takeoff. Both were unfortunate incidents, to be sure, but the subsequent disappearance of a manned two-engine J2C-12 in the Hindu Kush Mountains near Kabul took the cake.

The J2 took off with six passengers on a classified mission. The crew neglected to check in with the tactical military controller after becoming airborne, as procedure required. They were never heard from again. An A-10 pilot in the vicinity thought he detected wreckage in the mountains. He noted the coordinates. A short time later, in the middle of the Afghan night, I took to the air on an ad hoc mission to get a closer look at the wreckage.

I came up empty-handed after hours combing rugged mountains blanketed in snow and cloaked in darkness. The "wreckage" that the A-10 spotted was apparently nothing but the pieces of an old barn scattered about by a storm.

At daybreak, Combat Search & Rescue launched a round-the-clock search with helicopters and A-10 "Sandys." The A-10s were in charge of assigning search areas. They covered large blocks of terrain while helicopters rode low to scrutinize hills and streams and canyons. Since the Sandys had difficulty communicating with the low-flying choppers, I orbited above the lead Blackhawk to relay instructions from the jets. It wasn't overly efficient, but we made it work.

Just looking down upon that cold, barren land made me shiver in my warm GCS in Nevada. Somewhere down there in a mountain range the size of Alaska, six Americans may have survived a horrendous crash. Now they faced surviving the cold and wind and elevation. They couldn't last long, especially if they were injured. When a Predator went down, the United States lost four million dollars. When a manned flight crashed, we lost *lives*.

A *Beetle Bailey* comic strip may have contrasted the difference best. In the first panel, Sarge, Killer, and Beetle are watching a Predator fly above the landscape. "Those drones are great," Sarge exclaims. "They drop bombs with no pilots to get hurt."

"What if the enemy gets their own drones?" Killer asks in the second panel.

To which Beetle in his inimitable style replies, "Perfect. Then the drones can fight it out and we can all go home."

Were war that simple.

We scoured the mountains for days trying to find the downed airmen and came up with nothing. It was as if the airplane and its crew and passengers vanished into some Far Eastern Devil's Triangle. The incident never made headlines in the States. It wouldn't have even if we had located and rescued survivors. After all, none of the six or their aircraft existed in any official capacity. Special Operations knew the routine. I didn't even know their names.

It was official. Or at least as official as it was going to get until I had actual orders in hand. I broke the news to Trish: I would be shipping out to Iraq after New Year's. She didn't have much to say. I knew she didn't want to distress me by making a big deal of it, but I could tell it disturbed her.

I had expected to pull a tour at the Balad Air Base, launching and recovering Predators for pilots at Nellis. Instead, my orders

directed me, along with sensor operator Senior Airman Steve
Yob, to report to Ali Air Base, located about 180 miles southeast
of Baghdad. It was the only base between the capital and Kuwait.
There, instead of flying Predators, we would act as consultants
for the Italians who had recently acquired their own RPAs and
needed American expertise to teach them to fly in combat. I felt
flattered that I had been selected.

I was both nervous and excited. Not that I would be out
patrolling or engaging in firefights and suffering IEDs to explode
underneath my vehicle. I would be, more or less, among the rear-
echelon types that front-line grunts called "Fobbits"—those who
remained at the forward operating base (FOB) while everyone
else went out on patrol. Nonetheless, this would be my first time
actually setting foot in a war zone. You couldn't really count flying
Predator from Nevada or navigating RC-135s out of Crete earlier
in the war as "boots on the ground." Besides, I kept hearing horror
stories about attacks on American bases and guys being killed
and maimed by mortars and rockets. It seemed I would be taking
many of the same risks as other soldiers in combat. For the first
time since the war began, I felt a direct kinship with the troops I
had been supporting for so long from a distance.

As "the Professor," I was also looking forward to learning
more about the Muslim culture and our enemy, the jihadists.
Robert McNamara had demonstrated how important it was to
understand an enemy's motivation in order to defeat him. The
jihadist mindset was difficult for a Westerner to comprehend,
especially from a distance. I found myself baffled by the rational-
ization behind jihadist behavior and perspective. All that stuff
about a martyr's being awarded seventy-two virgins in Paradise
if he strapped on a Baghdad vest and committed suicide while
blowing up a busload of school children.

In an interview over Saudi Arabia's Al-Majd TV, a Saudi cleric
went so far as to promise prospective martyrs that their virgins
would be white. Very white. "Allah said that the black-eyed
virgins are beautiful white young women . . . whose skin is so

delicate and bright it causes confusion. . . . They are like precious gems and pearls in their splendor, their clarity, their purity, and their whiteness."

If that wouldn't make you want to go out and blow yourself up, what would?

Was there some kind of separate afterlife for pious but unwhite Muslim women, for women who weren't virgins? What kind of incentive had a Muslim girl to remain a virgin if her reward consisted of bevying up to some scroungy fanatic who kept shouting *Allahu Akbar* and blowing himself up? No wonder women who strayed had to be stoned to death; that was the only way to keep them in line.

Some of the messages put forth by the radical thinkers were downright baffling. Take the cleric Sheikh Muhammad al-Munajid, who issued a fatwa against "one of Satan's soldiers." Mickey Mouse.

Understanding all this definitely posed a challenge.

In the meantime, my final two weeks flying out of Nellis hit a dry spell. Fallujah II was winding down. Marines were moving house to house rooting out stray enemy holdouts. Harrowing enough for those on the ground, to be sure, but a real yawner from the air. Some days it was all I could do to stay awake for my shift in the seat.

It wasn't much better in Afghanistan. Karzai's inauguration as president, the first of the country's leaders ever voted into office, brought Vice President Dick Cheney and Secretary of Defense Donald Rumsfeld to the capital. That meant heavy security above and below. Any suspected Taliban or al Qaeda who so much as stopped on his way through Kabul was rousted and thrown into the clink. It would have been a good day for martyrs—except none dared show up. Besides, Afghan jihadists seemed to be more rational than those in Iraq. I flew overhead and looked down and not a dissenting voice did I hear.

Things picked up a bit on my last full shift at Nellis. On request of SpecOps, I was watching a house on the outskirts of Baghdad, covering for a pending raid. Like many Iraqi dwellings, its roof was flat with a stairwell leading up to it. A man kept popping onto the roof, looking about nervously, then ducking back inside. Like a prairie dog in and out of its hole watching for coyotes.

About fifty U.S. Army troops in Bradleys roared up the street toward the house. My "prairie dog" made another nervous check from the roof, saw them coming, and decided to run for it. He bolted down the exterior stairs, in one door of the house, out the back door, and over his neighbor's courtyard fence. He was hauling coal. Wiley Coyote could never have caught up with him.

But I could. I watched with some amusement as he pounded down a side street, looking back over his shoulder. A stray dog latched on to him, snapping at his heels. The dog chased him all the way to the mosque on the corner. The guy darted inside and slammed the door against the dog, who sat down on his haunches, threw back his head, and began barking and howling. *Here he is, here he is!*

I passed on the information to our guys on the ground. Bradleys and troopers surrounded the mosque. The outgunned dog took one look, tucked his tail between his legs, and headed for neutral territory. It was against U.S. ROEs to attack a mosque, but entering it was allowed. A couple of soldiers went inside, dragged the fugitive out with his wrists flexicuffed behind his back, and tossed him into a Humvee. I assumed he was on his way to the detention facility at Abu Ghraib.

Not a bad day's work. Another insurgent out of circulation. The only problem was that it seemed for every guy we caught, another fifty remained out there still causing mischief.

The shift ended. I went around the POC shaking hands with the MCC, analysts, other pilots, and the rest of the crews—people who had been my friends and work mates for the past months. Some were envious that Airman Yob and I were on our way to Iraq, the "Sandbox," while they stayed behind. When I stepped outside into a crisp but sunny winter day, I felt calm and controlled. This was exactly what I had wanted ever since Crete and the day the 173rd Airborne parachuted onto Harir Airfield.

My next stop was Iraq.

PART II

Ali Air
Base, Iraq

CHAPTER 14

Getting Real

A S AN RPA PILOT, I realized from the day I laid eyes on the Predator at Creech Air Force Base that I was in the vanguard of technology that was revolutionizing warfare. I was making history, not merely observing or studying it. When U.S. forces invaded Iraq in 2003 and I was on Crete, we had no RPA units in-country. We did have a couple of unarmed RQ-1s flying recce out of Kuwait, but that was it.

By the beginning of 2005, when I received orders to set actual foot in Iraq, we had 150 various unmanned vehicles in-country, including, of course, Predator. That number was expected to exceed two thousand by the end of the year. Defense Department people whose job it was to plan future warfare were already contemplating a country waging war without putting its own soldiers or civilian population in the line of fire. Call it "pain-free" military action. Except, of course, not for the enemy.

I had come of age on *Star Wars* and *Star Trek* reruns, and the modern versions of *War of the Worlds*, all far-fetched science fiction of the time but no longer so improbable in the twenty-first century. I could imagine swarms of air and ground robots seizing control of airspace and enemy cities while amphibious automatons made Iwo Jima–like beach landings to knock out pockets of resistance. Either under actual construction, in prototype stages, or already on the ground were twenty-two different unmanned systems ranging from flying RPAs with wings the length of a football field that could stay airborne for weeks to insect-sized and insect-appearing nano-robots whose parts consisted of single molecules. A Pentagon planning paper referred to as *Joint Vision 2020* predicted that one third of all U.S. combat aircraft would be unmanned by the end of that year. Ground and rear forces would also rely on robots, as would the navy with innovations such as robotic submarines.

Predator was the most familiar unmanned aircraft, as well as the most successful. A number of articles about it had appeared in various periodicals, along with TV features on the Military Channel and the Discovery Channel. Almost everyone, it seemed, had heard the story of how a CIA aircraft had reportedly fired a missile at Osama bin Laden and, regrettably, missed. Less recognizable to the general public were other unmanned aircraft systems that ranged from a Raven or a Wasp—small enough for an individual soldier to toss into the air like a model airplane in order to take a look above nearby rooftops, at the other side of a hill, or into the next street or alley—to the forty-four-foot-long jet-powered Global Hawk, which flew at sixty thousand feet altitude and could remain airborne for more than thirty-five hours at a time. In the works was the X-45C UAV bristling with precision-guided weapons.

In addition to unmanned machines that flew, a number of robotic configurations were also either being used or were almost ready for deployment by infantry on the ground. The Talon, for example, was initially built as an explosive ordnance disposal machine to handle dangerous unexploded ordnance by remote control. It had since been modified into at least two other applications—the modular advanced robotic system and the special weapons observation reconnaissance detection system, each of which could be fitted with machine guns or other weapons fired by a distant operator.

The forty-two-pound PackBot was one of the more amazing and utilitarian ground robots. About the size of a push lawn mower, it moved by using four independent flipper-shaped treads. Mounted with an assortment of cameras and sensors, the robot could move forward or backward, rotate on its axis, climb stairs, rumble over rocks, sneak down twisting tunnels, and even drive underwater. PackBot's smaller cousin, the MarcBot, about the size of a toy truck, had a video camera mounted on its tiny mast. It was the first ground robot to draw blood in Iraq after troops fighting in the streets of Baghdad and Fallujah discovered that they could jerry-rig a Claymore antipersonnel mine to it and send it snooping into a waiting ambush or into an alley or a building to blow up the bad guys.

The long-term goal of such technology, said a spokesman for Defense Advanced Research Projects, "is to create chips that reason and adapt, enable smarter sensors and achieve human-like performance." In other words, the U.S. military hopes to dehumanize military operations by creating machines that are more humanlike in performance.

"The trend," said an army colonel, "will be robots reacting to robot attacks. [Eventually] there won't be any time in it for humans."

Maybe so, but the Predator was futuristic enough for my fellow RPA pilots and me in 2004 and 2005. Few of us considered it to be a robot. To us, it was as much a warplane as the F-18 or the A-10, a mere extension of current airpower that required a true aviator to employ it in an effective manner. It had automatic systems, but it was not autonomous. If you wanted to squeeze out of it every ounce of performance, if you wanted to employ its weapons with the most effectiveness, then the Predator required a pilot and crew. That meant stick-and-rudder skills when the computer couldn't keep up. That entailed flying it and landing it by the seat of your pants the same as with any other airplane. That meant real human beings.

And that meant, also, that some of us had to go forward into the war zone to take off and land it for pilots back at Nellis. Or to train others how to do it. It was now my turn.

***

"Going to war" took over a week. Senior Airman Steve Yob and I booked commercial flights from Las Vegas to Atlanta and then on to Fort Walton Beach, Florida, where we boarded a "Rotator" DC-10 operated by Omni Air International Charter Service. After more than forty-eight hours en route, we reached Frankfurt, Germany, and camped out on a concrete floor for two hours while the DC-10 refueled and restocked. Another long hop brought us to Qatar and Al-Udeid Air Base, the last stop before we entered Iraq.

Al-Udeid, which boasted the longest runway in the region, sprawled across a giant swath of flat desert twenty miles south of the Qatari capital of Doha. With a total national population of only about 750,000, Qatar considered the United States to be its primary protector. In 1996, it built Al-Udeid at a cost of more than a billion dollars to encourage the U.S. military to base its aircraft there. In 2003, the United States began shifting its major air operation center for the Middle East from Prince Sultan Air Base, near Riyadh, Saudi Arabia, to Qatar. "Build it and they will come" was apparently a philosophy that worked for the tiny nation on the Persian Gulf.

For the rest of the week, Yob and I hung around Al-Udeid, living in a 288-tent complex called "Camp Andy" while we received daily briefings from the Combined Air and Space Operations Center (CAOC) on our consulting job with the Italians. The CAOC was the hub for all aspects of intelligence, operations, planning, and execution of airpower for wars in Iraq and Afghanistan.

The operations staff at the CAOC placed great emphasis on our having Italy's Predator and pilots up and flying ahead of Iraq's legislative elections on January 30, 2005, in order to enhance surveillance in the Italian military–controlled south province of Dhi Qar. Choosing representatives for the newly formed and transitional 275-member Iraqi National Assembly, the first general election since the 2003 invasion, marked an important step in turning control of the country back to its own people. Coalition forces were naturally apprehensive; a lot was riding on the elections being peaceful. Predators, both Italian and American, would play a major role in surveillance protection. Abu Musab al-Zarqawi and other leaders of the insurgency were threatening major actions to disrupt voting.

"We have declared a fierce war on this evil principle of democracy and those who follow this wrong ideology," Zarqawi declared through a terrorist website. "Anyone who tries to help set up this system is part of it."

The small Italian contingent in Iraq had had a rough time of it. Back in May 2004, when the cleric Muqtada al-Sadr was showing himself in Sadr City, Fallujah, and elsewhere, his Mahdi Army drove the Italians from their main base in the southern city of An Nasiriyah with grenade and mortar fire, wounding at least ten soldiers. Nasiriyah had become famous at the beginning of the war when the U.S. 507th Maintenance Company took a wrong turn and was almost wiped out in the city. Eleven soldiers were killed and six captured, including Jessica Lynch. Lynch and four others were eventually rescued.

Fleeing Nasiriyah, the Italian force of about three hundred, under the command of Col. Antonio Albanese, relocated to Tallil Air Base, seven miles away on the south side of the Euphrates River. Since then, the Italian air force had acquired one Predator system that included five aircraft, all of which were RQ-1s, a generation behind the MQ-1s that we were flying out of Nellis. The RQ-1 was not satellite capable, which meant that Yob and I would be training pilots using line-of-sight radio links that limited operations to within a hundred miles of the base.

"You have to have the Italians operational by at least January 28." That was drilled into us during the CAOC briefings.

The C-130 Hercules transport was packed with excited soldiers bound for Tallil Air Base, recently renamed as Ali Air Base as part of an effort to remove the reminders of Saddam's rule. Most of them were kids. At least they were kids compared to my wizened thirty-four years. Many of them were not much out of high school, and some had not even been out of their own hometowns before they enlisted. I suspected they would look a lot older by the time they saw their share of Iraq and returned stateside.

Yob and I were geared for action—Kevlar helmets, flak jackets, chemical warfare MOPP (mission-oriented protective posture) suits, 9mm pistols strapped to our hips. With an exaggerated sigh, Yob let himself fall into the web seating next to me. He grinned impishly as the C-130 lifted off from Qatar.

"I hope the Italians aren't in charge of the defenses at Ali," he quipped. "Did you know they sew extra material into the armpits of their uniforms to make it easier to surrender?"

"You're thinking of the French," I said.

"Oh, yeah," he said.

As the Hercules entered Iraqi airspace, the loadmaster standing forward by the steps leading up to the pilots' cabin turned the interior lights to red to attract our attention.

"Make sure your seat belts are secure," he counseled over the intercom. "In the event the airplane is engaged, we don't want to toss you around like pebbles in a tin can during evasive maneuvers."

"This is a lot realer than flying Predator, Major," Yob observed. I had been promoted from captain to major the week before we left the States.

"I suspect things are about to get even realer," I said.

We couldn't see out of the plane because there are no windows near the troop seats of a C-130. All I knew was that it was dark. The transport screwed itself down to the airport in a spiral approach, then dropped in so fast and steep that some of the GIs let out little whoops to recapture their stomachs.

The back ramp dropped after the touchdown and rollout. Troops grabbed their gear and piled out onto the parking apron in front of a long, low building with its windows blacked out. I assumed it was Ops. From what I could see, very few lights burned anywhere. Even the runway lights extinguished as soon as the plane landed and taxied off.

Welcome to Iraq. A grayish brown powdery dust covered

everything that did not move. I tasted it in my throat, felt it gummy with moisture in the corners of my eyes. Two American facilitators were waiting to help Yob and me with in-processing, after which we hopped into a waiting pickup truck and sped off in search of the Italian compound where we were expected to bunk.

Ali occupied nearly twenty square miles of desert and encompassed within its perimeter two runways; thousands of airmen, soldiers, and civilians from at least three coalition nations; and the ancient ruins of Ur, Iraq's most famous archaeological site and probably the first real city in the world. Even in the darkness, I spotted abandoned military buildings with windows shot out and walls pocked from small-arms fire. Eroded craters here and there marked where bombs had been delivered by American tanks or aircraft.

Ali had been a fighter base before the 1991 Gulf War. The United States bombed it repeatedly during Desert Storm. By March 2003 and the launching of Operation Iraqi Freedom, it was an air base in name only. Although it still housed air defense systems, Iraqis had not flown aircraft out of it in more than a dozen years. Runways, taxiways, and ramps had not been maintained. Trucks hauled in water. The only available electrical power came from a system of portable generators and batteries.

After the base fell to coalition forces in 2003, the 407th Expeditionary Civil Engineer Squadron immediately hauled in nearly ten thousand truckloads of fill dirt, assembled 350,000 square feet of tents and portable wooden buildings, trenched forty thousand feet of electrical cable, and buried more than five miles of underground water pipe to make the base once more functional. Although still desert for the most part, isolated from any significant civilian population, it was again an important military airfield along a major supply route from the south.

I wasn't much impressed by what I had seen of it so far. We finally found Italian ground after driving around in black desert for quite some time. "The scenic route," our guide joked halfheartedly. A gate guard who spoke almost no English knew nothing about the Predator and its personnel. He pointed off toward the main side of the base and shrugged his shoulders.

Our driver turned the truck around and headed back the way we had come. Somewhere in the distance, security troops traded machine-gun fire with guerrillas probing the perimeter, the chatter faint in the wind. Yob seemed concerned.

"I hope the Italians don't give up before we get there," he remarked with a stab at his old humor.

Unbeknownst to Yob and me, we inadvertently ventured out of the controlled sector of the base and into no man's land for a mile or so in our search for the Italian tent city. We didn't find out about it until days later, which caused Yob to roll his eyes and mutter to himself.

The Italians were blacked out and slumbering when we reached the bivouac. Some of the tents emitted faint snoring. A sentry awakened a captain, who hadn't expected us until the next day.

"No worries," he assured us. "You may—how you say?—*crash* in my tent with me."

CHAPTER 15

In Its Own Time

"YOU ARE REQUIRE TO carry a weapon at all time. It is most danger on the perimeter," Lt. Col. Antonio Gentile said in good but fractured English. A short, thin man with a big nose and even bigger ears, he was the Italian Predator squadron commander and one of three pilots detailed for Yob and me to train.

The other two pilots were Lt. Rafael Orsolini and Capt. Riccardo Venuti. Yob and I met all three the morning following our arrival during a leisurely breakfast at the Italian mess tent. The younger officers seemed disciplined, focused, and eager to get started. Although Colonel Gentile expressed equal enthusiasm, he was more laid back, Latin style. Like many of his countrymen, he possessed a different work ethic than Americans. His subordinates followed his example. It was after nine o'clock in the morning when we went for breakfast. However, we discovered, they supposedly worked late to make up for their relaxed breakfast and post-meal custom of sitting around sipping strong coffee and chatting. I could readily adopt that ritual.

"Everything is to be in its own time," Colonel Gentile said with a smile of tolerance for rowdy, rush-crazy Americans. "It is to say more civilized."

Yob hooked his M16 rifle over the back of his chair and stretched his long legs leisurely out to one side of the table, a coffee cup hooked on his finger. The self-satisfied smirk he wore told me he could get used to the Italian way as well.

"I like *civilized*," he said, making the Italian officers laugh.

There was, in my opinion, *nothing* civilized about making war. War was what happened when civilization broke down. But I kept my mouth shut and my opinion to myself. Sometimes it was tough being "the Professor" and a career warrior wrapped in the same package.

The five of us and Maj. James "Rainman" Hoffman, our liaison to American headBobbys, finally reached the Italian operations center shortly before noon. Ops was a remodeled underground former Iraqi bunker about the size of a baseball diamond. It would eventually house the Italian equivalent of our POC, along with a communications center shared with an adjoining helicopter detachment. I was surprised, however, considering the January 28 deadline to get the pilots up and flying, to find that little had been done toward making the center operational.

The GCS, the ground support equipment, most of the maintenance equipment, and even the two RQ-1 Predators presently in-country were still in packing crates. The Italians walked around the Predator "coffins," patting them as though they were the bellies of expectant mothers.

"Is it true, as we have heard?" Lieutenant Orso, short for Orsolini, asked. He spoke excellent English, having attended air force flight school in the United States. "At the beginning of the war, you Americans used a Predator to fire a Hellfire missile that knocked out Crazy Baghdad Bob's Iraqi TV satellite antennae near the Grand Mosque in Baghdad?"

"That's the rumor," I conceded.

The shorter of the two runways would be ours to use, only a short taxi away from the hangar and the POC. My fear was that we wouldn't be able to use *anything* unless the Italians got off their behinds, at least not in time to be airborne before the Iraqi elections. Colonel Gentile seemed unfazed by my concerns. It would all be accomplished, he assured me—everything unpacked, the communications tower erected, the aircraft and GCS assembled, and the POC designed and made functional.

"Everything is to be—" the colonel began.

"I know, sir," I said with a tight grin. "Everything in its own time."

Ed Kimzey, a civilian instructor pilot from General Atomics Aeronautical Systems, manufacturer of the RQ-1s, had accompanied the Italian pilots to Iraq. He looked more like an auto mechanic than a pilot.

"Don't worry," he reassured me. "I've worked with Italians. They'll get it done—"

We both laughed and completed the sentence in unison, "—in its own time."

Although Kimzey would be primary flight instructor, he wasn't a combat instructor. Therefore, Yob and I would serve as mission instructors to make sure the pilots knew how to employ Predator as a reconnaissance platform. We didn't have to be concerned with weapons; the Italian aircraft would fly unarmed. Additionally, part of our job was to act as liaison in integrating Predator ops into the coalition air operations, making sure that our sorties were programmed into the daily air tasking order and that Colonel Gentile and his men knew the rules that governed flights in Iraq.

True to his word, the colonel quickly got everything organized. The unmanned Italian air force in Iraq began to take shape. In the meantime, since Yob and I had little to do, we nosed around base and oriented ourselves with the pace and rhythm of life on a combat outpost in a war zone. I soon learned that the Italians weren't alone in launching a new start-up air operation at Ali.

The U.S. government had given the Iraqi Air Force three C-130 transports to use for hauling cargo in the name of Iraqi Freedom. Nobody was supposed to know that Iraqis were flying them. If word leaked out, the air base would become a hot target for insurgents, and members of the Iraqi Air Force would have big X's on their backs.

One evening I met one of the Iraqi airmen while we were both using the laundry tent. My Arabic was all but nonexistent, limited to *sabah el khayr* ("Good morning") and *shokran* ("thank you"). His English was somewhat better, good enough for him to explain that Saddam Hussein had prohibited anyone from learning English because he believed that that knowledge would make it easier for dissidents to flee the country. He had been an avionics technician for Saddam's air force. Any trips he made out of the country as part of his duties were closely monitored. Family members were required to report to the government while he was gone. They would be killed if he failed to return.

His family, he said, was still at risk, only not from Saddam. Terrorists and insurgents would murder him and his family if they learned he was working for the new air force.

They wouldn't learn it from me. We shook hands. What a harrowing life he must lead.

Life at war took some getting used to, what with everyone on base packing heat and swaggering around like John Wayne. To me, the new guy in town, the atmosphere was a lot like the Old West must have been as depicted in the movies. I kept expecting the social order to break down at any time. The American chow hall especially was a circus of exaggerated machismo. United States troops mixed with Dutch and Italians, civilian contractors, and mysterious types in unmarked uniforms I couldn't identify, all waiting with rifles slung across their backs and pistols on their hips for the pasta line, the pizza line, a short-order line, or meat and potatoes and ice cream.

After a few days, the 9mm on my hip began to feel almost natural.

"Matt!" Colonel Gentile exclaimed, as though he had almost forgotten the hour. "We must not tarry. It is—how you say?— *Miller Time!*"

Little wonder it had taken six days simply to get the equipment set up and the POC and airplanes ready to go. All those late breakfasts with coffee and conversation, followed by lunches equally as long. Contrary to the Latin stereotype of a nap in the afternoon compensated for by working late, the colonel and his pilots were ready to call it a day by 1800 hours. *Miller Time.* After chow, laughing, and chattering, we made our way for grappa and more coffee to a little café that the Italians had set up. There, we caroused much of the night away until, arm-in-arm-in-arm, we made our way to dusty tents for a few hours' sleep.

"Simply because there is war is no call to suspend civilized conduct" was Colonel Gentile's impeccable reasoning.

Who was I to argue with such logic?

"Reckon I could transfer from the U.S. Air Force to the Italian air force?" Airman Yob wondered, sounding at least half serious.

Even as uptight and responsible as I knew myself to be, I could have grown accustomed to the laid-back Italian attitude. I had to keep reminding myself, as well as Colonel Gentile, that time was running out if we hoped to have Predator in the air by the elections. I counted down the days. Twelve, eleven . . .

Ed Kimzey, the contract flight instructor, had done a fantastic job training the pilots in technical proficiency. Apparently, according to Kimzey, they could fly the aircraft okay. I didn't know one way or the other; we hadn't flown one yet. We had to get them in the air before Yob and I could carry out our part of the bargain. The pilots needed actual operational and tactical combat experience. I began to understand why Mister Moran during my own training at Creech often lost his cool and yelled. I had to suppress the urge to do likewise at times. *What the fuck's the matter with you? Got your head up your ass?*

I never used that type of language. If I so much as frowned, Colonel Gentile, with that big Latin grin of his, polite and thoughtful to the point of being deferential, was right there buying Yob and me coffee and thanking us for all we were doing to get them in the air. In contrast, the CAOC seemed remote and less than enthusiastic about supporting the mission, more or less ignoring us. It seemed that Yob and I were on our own, sink or swim, as though our teaching the Italians to operate RPAs interfered with other operations that we should have been doing. I resented it, especially the delays when I requested some sort of support, but there was little I could do. My take on the situation was that this was going to be a long and ugly war and the United States needed all the help we could get, even from the easygoing Italians.

Ten days before elections, we reached and passed a major milestone in developing an Italian remote pilot combat reconnaissance capability: we taxied. Flying line of sight meant no dedicated LRE teams; pilots had to take off and land their own aircraft. This taxi test would make sure not just the aircraft, but the datalink and the control equipment in the GCS was ready to go prior to the first flight.

They insisted on first having their usual two-hour lunch. It was almost sundown by the time we got started. Colonel Gentile, as befitted his rank and position as squadron commander, went first. My heart leapt into my throat as Mr. Kimzey and I hovered behind him and his sensor, watching the main video screen intently as the sleek little airplane lumbered down the taxiway into a red setting sun, then turned around and came back to give Captain Venuti a try at it.

"Time to knock it off!" I shouted with relief after the three of them had practiced until nearly full dark without putting the plane in a ditch. "Tomorrow, you take off."

They stood up and cheered like schoolboys. You couldn't help but love the irreducible Italians.

I didn't sleep well that night. Flying around Iraq seeking targets was a complicated business. I wasn't sure that Gentile and his pilots and sensors had yet acquired the proper skills for it. But with only nine days to go, it was time to throw them into the deep end.

The Predator was fresh out of the box and required a functional flight check flight prior to the first combat mission to make sure everything worked. Kimzey and I helped with that, after which Venuti took over a plane to become the first to successfully get airborne. Orso and he took turns for the rest of the day flying Predator out over ancient Ur and to the outskirts of Nasiriyah as they put the aircraft through her paces. They were ecstatic and promised a big party that night to celebrate their achievement.

War has a way of intruding and putting a damper on the best of moods. Yob and I with our new friends departed the new Italian POC after a successful day of practice flying, all of us in high spirits and anticipating not only a raucous night at the café but also a real mission within the next day or so. Colonel Gentile's executive officer met us outside the GCS. The look on his face said the news he bore wasn't good. I hadn't picked up much Italian in the two weeks since we had arrived, but I noticed how all levity escaped the Italians like air from a balloon.

"What's up?" I inquired anxiously.

Captain Venuti translated the message—first the news about the Americans. Insurgents had ambushed a U.S. Army convoy passing through Nasiriyah. Our guys had killed several of the enemy, but IEDs and hostile fire wounded four of our soldiers, one critically. They were taken to the Ali Air Base hospital, the U.S. Air Force's largest and most well equipped emergency care medical facility in this part of Iraq. All four were expected to survive.

The Italians hadn't fared so well. One of their helicopters returning to base after a patrol over the city took ground fire—a single shot from some insurgent hiding in a wadi. The Hajji was either lucky or the best shooter in the country. The bullet struck the door gunner underneath his right arm, where there was a gap in his body armor, killing him instantly.

Lieutenant Orso's eyes brimmed with tears. "We all knew him," he said.

An honor guard in front of headBobbys lowered the Italian flag to half-mast, the universal sign of mourning for a fallen warrior.

The colonel and his two pilots went to attention and snapped off salutes. Yob and I did likewise. Desert wind nipped at the legs of our flight suits, sand hissed across a tin roof, and tent canvas popped like distant gunfire.

"We will carry on," vowed Captain Venuti. "With the Predator, perhaps we are saving lives of our men."

I was no longer a spectator from 7,500 miles away. You couldn't get much closer to the war than this.

CHAPTER 16

Elections Iraqi Style

ONE WEEK UNTIL ELECTIONS—AND the Italians had yet to complete an actual mission. I had high hopes for today, but severe interference with the radio link and a malfunctioning ground antenna forced us to abort and return the Predator to base shortly after it took off.

Six days to go. A two-hour lunch got in the way. The sun was already going down by the time we received mission briefing and everything was ready to fly. The Italian commanding general present at the GCS for this momentous event stood in the doorway and gazed out across the runway tinged red by the dying Iraqi sun.

"*Domani,*" he decided. "Tomorrow."

An American GCS was equipped with an extra computer not present at the Italian station. This additional screen, normally located between the pilot and sensor operator, displayed a map of airspaces established to conduct the air war. Without the screen and its map, a Predator pilot might inadvertently blunder into somebody else's fight and get shot down. Absence of the screen also made it difficult to locate targets.

The Italian GCS was wired for the computer, but no one had installed it. Colonel Gentile had been bugging his communications team for days to get to it, but everyone in his sector of Ali Air

Base seemed to live by the same philosophy: *Domani. Everything in its own time.*

There came a point when *when in Rome* ceased to suffice. "We'll hook it up ourselves," I finally decided.

Colonel Gentile and his pilots beamed. "How can we be of help?"

"I'll need a long monitor cable and a flat-panel computer screen. I'll dig up the cable if you can beg, borrow, or steal the screen."

I drove a truck to the American sector and negotiated with the U.S. Air Force NCO in charge of the computer shop. Our communications tent alone was larger than all the Italian space combined. The air force sergeant and his men cobbled and spliced together sections of leftover cable. It looked like it would work. All it cost me was a promise that the sergeant could come over to the GCS and watch a mission as soon as we were able to launch one.

In the meantime, Colonel Gentile took me literally when I advised him to "beg, borrow, or steal" a monitor. I didn't ask him how he procured it, but I doubted that the monitor was Italian. We propped the screen on a chair, connected the cable—and it worked.

"It is fabled American ingenuity," Orso declared.

That was the fifth day before the polls opened.

Four days to go. A minor dust storm in the morning cut visibility to almost nothing, forcing us to scrub scheduled flights. *That* I couldn't blame on the Italians. Since we couldn't fly, I didn't feel nearly so guilty about our customary two-hour lunch followed by five gallons of coffee. Lieutenant Orso glanced outside when someone flung open the tent flap to leave.

"The weather, she is clearing," he triumphed. He was the more eager of the three to fly a real mission.

Everyone jumped up and hauled for the GCS. The weather, the plane—everything was a go. Yob took up his position behind the sensor operator while Colonel Gentile deferred to Captain Venuti to pilot the takeoff and initial leg of the mission.

Their first job was a simple one: fly sixty miles north, spy on a couple of suspected insurgent compounds, then return to the base with the video and related intelligence.

As excited as children at Christmas over their first real-world undertaking, the three pilots took turns at the controls while I stood behind them with a running patter on tactics and techniques—how to position the aircraft to obtain the best view of a target; how to prevent being detected through orbiting against the sun and staying downwind to muffle engine sounds; what behavior to look for in a bad guy.

There was a lot they still had to learn and practice. We needed work on approaching a target, setting up for the best video quality, directing troops into a raid, IED searches, convoy route clearance, tailing cars and other vehicles through busy traffic. Confidence in my students was growing.

But for today, it was enough to launch and examine an actual target. Although the run generated little in the way of actionable intelligence, it was a good drill for the pilots in receiving input from a supported unit, in this instance an Italian special forces team, acting on that input, sending data and receiving feedback, all without stepping on one another and clotting up the communications net.

The Italians were thrilled. We had become operational in time for the elections, with three days to spare. As soon as they landed the Predator, the five of us ran out and triumphantly dug out the American and Italian flags that we had stashed aboard the plane for its maiden voyage. With flags waving, Colonel Gentile popped the cork on a bottle of champagne that he had been saving for the occasion. Everybody hosed down everybody else with the bubbly.

When in Rome . . .

My job was almost finished at Ali Air Base. Yob and I had the Italian Predators and crews up and on-line. In a few more days, almost immediately after the elections, we would be receiving orders to redeploy stateside. I was ready to get out. At least I wasn't stuck here for a year's tour like the poor bastards in the army. I couldn't seem to stay well in this climate. The air was thick with smoke, dust, and exhaust fumes. I sweated my ass off during the day and froze what was left of it at night. I developed respiratory problems my first week in-country and went around coughing and wheezing. The Italians seemed to think all ailments could be cured by prodigious quantities of grappa, but so far it hadn't worked.

There was little ventilation in the bunker that served as the Italian POC. The Italian pilots were kind and generous, but they had different hygiene standards than those of most Americans. It got pretty crowded in the GCS during extended-duration combat missions. Immediately after the Predator landed from a flight, Yob and I made hasty excuses to escape the bunker and get back into the relatively fresh air of smoke, dust, and exhaust fumes.

"My personal favorite fragrance is Colonel Gentile's," Yob joked. "It's a tantalizing blend of underarm BO with an essence of crotch and an enticing blend of halitosis. By comparison, Major Martin, you and I are a pair of weenies."

Not particularly amused, I turned around and coughed at him.

With only two days to go in the countdown to elections, I kept our allied pilots in the air for eight-hour flights over and around Nasiriyah. The city of about a half-million people would be Italy's primary responsibility when the polls opened. So far, it remained relatively quiet, at least in comparison to what I had grown accustomed to in Fallujah.

Flying Nasiriyah provided a study in contrasts between the ancient, the merely old, and the modern. Buildings in the "old town" section were constructed of sun-dried bricks not that much different from those used by the ancient people who built the archaic city of Ur thousands of years ago. Blowing sand had scoured structures and the mud walls that still surrounded the old part of town to the color of sand. Outside the walls rose a more modern finance and business district of glass and steel.

A Predator flies over a range in Nevada on September 6, 2007, while being filmed for an U.S. Air Force recruiting campaign. The vehicle is assigned to the 11th Reconnaissance Squadron, 432nd Air Expeditionary Wing, out of Creech Air Force Base, Nevada. *USAF photo by Master Sgt. Scott Reed*

A Predator test fires a Hellfire missile on the range in Nevada. Armed with two Hellfires, the Predator can fly at altitudes of 25,000 feet and remain airborne for up to twenty-four hours. *USAF photo*

Airman 1st Class Caleb Force (right), a sensor operator, assists Predator pilot 1st Lt. Jorden Smith in locating simulated targets during a training mission at Creech Air Force Base. Both are assigned to the 11th Reconnaissance Squadron. *USAF photo by Senior Airman Nadine Y. Barclay*

Airmen of the 432nd Aircraft Maintenance Squadron at Creech Air Force Base reassemble a Predator after it has returned from Afghanistan. The stencils on the side symbolize the number of missiles it fired in combat. *USAF photo by Senior Airman Larry E. Reid Jr.*

Then Major Matt Martin in front of a mural depicting Saddam Hussein at Ali Air Base in Iraq. Note bullet holes that have riddled Saddam's image. Ali, formerly named Tallil, used to be Saddam's air base. *Matt Martin*

American and Italian crews at Ali Air Base in Iraq (left to right): Major James Hoffman, American liaison officer; Ed Kimsche, civilian Predator contractor; Captain Riccardo Venuti, Italian pilot; then Major Matt Martin; Lieutenant Rafael Orsolini, Italian pilot; and Senior Airman Steve Yob, Martin's sensor operator. *Matt Martin*

The entire Italian Predator squadron poses with then Major Matt Martin (far right) following the squadron's first successful combat operational flight over Iraq. *Matt Martin*

A Predator armed with Hellfire missiles flies a combat mission over southern Afghanistan. The MQ-1 provided interdiction and armed reconnaissance against critical targets for Operation Enduring Freedom. *U.S. Air Force photo by Lt. Col. Leslie Pratt*

A mortar team of six insurgents near Balad Air Base, Iraq, targeted by a Predator crew using the infrared camera. Two individuals can be seen between the trees above the crosshairs. The on-screen caption identifies

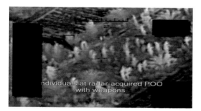

the insurgents as being at the mortar point-of-origin (POO) along with their mortar tube. Classified targeting information has been blacked out. *Released by the Multi-National Corps-Iraq, Sep. 2005*

The explosion of a Hellfire missile against the mortar team. "LRD LASE DES" indicates that the laser from the Predator targeting pod is still firing to guide the missile to its target. Several trees can be seen being

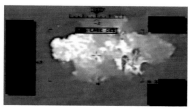

blown back by the explosion. Five insurgents were killed. The remaining insurgent was captured by coalition forces. *Released by the Multi-National Corps-Iraq, Sep. 2005*

This daylight view of a parking lot at Creech AFB, shot from ten thousand feet during a training mission and displayed on a computer screen, demonstrates what the Predator sees through its MTS targeting pod camera. *USAF photo*

Major Matt Martin with *Ruby*, the Predator he named for his wife during his tour at Balad Air Base. *Matt Martin*

The hardened aircraft shelter that housed the fleet of MQ-1 Predators at Balad Air Base, Iraq. Two line-of-sight antennae, used to control local aircraft, are affixed to the top of the shelter. *Matt Martin*

Crew chief Staff Sgt. Sean Pietre and Senior Airman Rothschild Pierre-Louis III, U.S. Air Force 46th Expeditionary Aerial Reconnaissance Squadron, load a Hellfire missle onto a Predator at Balad Air Base. *USAF photo by Staff Sgt. Cohen A. Young*

Then Brigadier General Frank Gorenc, commander of the 332d Air Expeditionary Wing, Balad Air Base. He was fascinated with the tactical use of UAVs and was, at the time, the highest-ranking officer to learn to pilot a Predator. He championed Major Martin's plan to provide a Predator base defense system for Balad. *Matt Martin*

United States Air Force 46th Expeditionary Aerial Reconnaissance Squadron pilots Capt. John "Disco" Songer and Airman 1st Class Stephanie L. "Princess" Schulte operate individual Predators at Balad Air Base in support of Operation Iraqi Freedom. *USAF photo by Staff Sgt. Cohen A. Young*

The team for Operation Neighborhood Watch, initiated by Major Martin (left, standing) to protect Balad Air Base. It was the first time the Predator was integrated in defense with U.S. Army ground forces and manned air force aircraft. Senior Airman Falisha Rexford (third from right, standing) and her pilot, Capt. Bobby Rangler (second from right, standing), were instrumental in the success of the program. Martin's sensor operator, Senior Airman Kimberly Joan (kneeling, right), was his right hand during combat operations. *Matt Martin*

Over the decades, Nasiriyah had experienced its share of violence and war. Britain conquered the city during World War I. At the time, it was a part of the Ottoman Empire. More than four hundred Brits and Indians and up to two thousand Turkish soldiers were slain in the battle that began on July 24, 1915. The years that followed saw an insurgence of political activity and movements in the city, including the founding of the Iraqi Ba'ath Party, which ultimately brought Saddam Hussein to power.

Nasiriyah marked the farthest point that the coalition forces reached during the 1991 Gulf War when the U.S. 82nd Airborne Division occupied the main road outside city limits. After American forces withdrew, the majority Shiite population in the city revolted against Saddam Hussein's Sunni rule. Saddam quashed the rebellion by massacring entire families. Thousands of refugees fled to other countries.

A dozen years later with the onslaught of the second Iraqi war, heavy fighting occurred in the city following the ambush of the U.S. 507th Maintenance Company of Jessica Lynch notoriety. The 2nd Marine Expeditionary Brigade crushed all resistance in less than a week. Other than a truck bomb that killed seventeen Italian soldiers and eleven civilians in November 2003, cleric al-Sadr's Mahdi Army uprising in April 2004, and the occasional IED or sniper attack, Nasiriyah had remained reasonably "pacified." Shiites cheered and marched through the streets laughing and dancing when U.S. soldiers captured Saddam Hussein hiding in a hole in the ground.

Everything could change on election day, however, what with Abu Musab al-Zarqawi threatening a "fierce war" against the "principle of democracy" that would "wash the streets in blood." Insurgents were getting ready. I picked up from the U.S. intelligence tent a translated article from an underground insurgent newsletter in which a Mujahidin leader commander ordered his followers, who had apparently gone home on pass, to report to their stations within three hours to renew actions against American infidels and their agents. *Praise and peace be upon Allah.*

Colonel Gentile and his crews showed up at the bunker carrying assault rifles and machine guns and wearing protective body armor from their ears to below their crotches.

Even the Americans geared up for a Vietnam-like Tet Offensive. Orders came down from the base commander for all troops, Fobbits and front-line troops alike, to draw full combat equipment—Kevlar helmets, body armor, rifles, extra ammo. I didn't sleep well the night before the voting booths opened. I half expected hordes of bloodthirsty foreign fighters to storm the base, as in *Lawrence of Arabia*. I stepped outside my tent in the middle of the night and surprised Yob gazing outboard toward the perimeter.

"I thought I heard machine guns," he explained.

We both listened.

"It kind of raises the stakes," he remarked. "I mean, actually *being* here."

The slide of my 9mm pistol had been sticking. I had taken it to the armory to have it checked out to make sure it was ready for action. Blasting some IED team with a Hellfire while I was in Nevada rather than confronting the prospect of having to shoot an enemy fighter point-blank with a sidearm did indeed, as Yob said, raise the stakes. I expected tomorrow to be the big test for my Italian pilots—and perhaps for Yob and me as well.

We launched a Predator at dawn, an hour before the voting booths opened, to fly a twelve-hour mission eighty miles in diameter centered on Nasiriyah. Long lines were already forming at schools, mosques, and other public polling places. People seemed solemn under the weight of their new responsibilities, while at the same time festive and celebratory. The voting began. The monitor screen showed people emerging with huge smiles and holding up index fingers stained with indelible blue ink from having been fingerprinted prior to voting.

Iraq cast a total of 8.4 million votes, a 58 percent turnout. Better than some presidential elections in the United States. Only

2 percent of the Sunnis went against Zarqawi's proclaimed boycott and showed up at the polls. As for the violence that Zarqawi threatened, election day in Iraq turned out to be relatively anticlimactic, without major disruptions.

Insurgents launched less than a hundred attacks nationwide, including nine suicide bombers. Of the forty people killed in the actions, thirty-six died in Baghdad, the work of six martyrs who blew themselves up in crowds of people waiting to get into polls. In addition, a British C-130 crashed near Baghdad, possibly as a result of hostile fire, killing most of the passengers and crew. By noon, it became clear that Tet was not going to happen. Those of us holed up in the Predator bunker at Ali began shedding hot and uncomfortable body armor. In retrospect, I felt a bit foolish at having shown up for duty in a secure bunker laden with weapons and combat gear.

From the time that polls opened until they closed, our Italian Predator over Nasiriyah scrutinized 158 Iraqi polling stations and scanned numerous highways, streets, roadblocks, power plants, broadcasting stations, and other buildings looking for suspicious activity such as men loading or unloading weapons, planting IEDs, setting up sniper positions, or assembling in preparation for an attack. The most exciting episode we observed was a couple of uniformed Iraqi cops dragging some guy out of a voters' line and pummeling the hell out of him on the way to their squad car.

"He probably deserved it," Yob commented.

In other parts of the world, including the United States, groups of protesters marched in support of the Sunni boycott and against the U.S. presence in Iraq. Zarqawi undoubtedly assumed that the marches were supporting *him*.

CHAPTER 17

Sunnis and Shiites

L IKE MOST WESTERNERS, I possessed little understanding of the differences between Shiite and Sunni Muslims, the two main sects of the religion, until I began studying Muslim history and culture. Muqtada al-Sadr was Shiite. Abu Musab al-Zarqawi was Sunni, as was Osama bin Laden. Al Qaeda was composed mostly of Sunnis. Hezbollah recruited Shiites.

In 1921, after World War I and the fall of the Ottoman Empire, Winston Churchill was said to have inquired as to the "religious character" of the leader he planned to install in Baghdad. "Is he Sunni with Shaih [Shiite] sympathies or a Shaih with Sunni sympathies? I always get mixed up between the two."

I daresay not many modern Western politicians knew the differences as Iraq threatened to explode with sectarian violence because of religious strife that went back more than fourteen centuries.

Both sects of Islam held that the Holy Book of the Quran, or Koran, was revealed to the prophet Muhammad, "peace be unto him," over a period of some twenty-three years beginning in A.D. 610, when Muhammad was forty years old, and ending in 632, at his death. Muslims regarded the Quran as a series of divine messages starting with those revealed to Adam and proceeding through Abraham on up to Muhammad, the last of the prophets. The

Muslim religion contended that the Quran was the main miracle of Muhammad and proof of his prophethood, having been delivered to him by the angel Gabriel in Mecca, Muhammad's birthplace. Mecca in Saudi Arabia has since been considered the center of the Muslim religion and its holiest site. Muhammad declared that no non-Muslim would ever be allowed in the city. Even when Mecca was modern and cosmopolitan, it was closed to non-Muslims.

The two branches of Islam diverged in a battle for the prophet's mantle after he died. Followers could not agree on whether to choose bloodline successors or to select leaders most likely to follow the tenets of the faith, whether they were Muhammad's direct descendants or not.

Shiites wanted to go the bloodline route. For them, Muhammad's grandson, Imam Hussein, was the rightful heir. However, Muhammad's father-in-law seized power for the Sunnis, representing for that faction the course it thought most likely to maintain the purity of Islam. When Hussein and his family refused to pledge their allegiance, the Sunnis besieged them at Karbala and slaughtered them and eighty-seven of their warriors.

Sunnis ruled the faith continuously in the Arab world until the breakup of the Ottoman Empire following the end of World War I. Heirs of the first four caliphs, though three of the four were not of Muhammad's bloodline, were recognized by Sunnis as legitimate religious leaders.

Shiites, on the other hand, believed that only heirs of the fourth caliph, Mahdi Ali, a direct descendent of Muhammad, were legitimate successors. When the Twelfth Imam, Muhammad Al-Mahdi, a Muhammad heir, mysteriously disappeared in 874 at the age of five, Shiite Muslims believed that they suffered the loss of divinely guided political leadership. Not until Iranian Ayatollah Khomeini in 1978 did the Shiites again live under what they considered to be legitimate religious authority.

When the Sunni-dominated Ottoman Empire collapsed to the British in World War I, the Sykes-Picot Treaty placed Iraq under the direct rule of France and Britain. Rogue Sunni military officers who assisted the British in overthrowing the Ottomans monopolized top political and military positions for decades, until the overthrow of Saddam Hussein in 2003.

Radical fundamentalist Muslim movements that spawned twentieth-century terrorism sprang out of the abolishment of the Ottoman caliphate. In 1928, four years after the Ottoman downfall, an Egyptian schoolteacher named Hasan al-Banna was reportedly appalled at European influence and "the wave of atheism and lewdness that engulfed" the Muslim world. Westerners, he raged, had "imported their half-naked women into these regions, together with their liquors, their theaters, their dance halls, their amusements, their stories, their newspapers, their novels, their whims, their silly games, and their vices." He was further incensed by European schools and scientific and cultural institutions that "cast doubt and heresy into the souls of Muslim sons and taught them how to demean themselves, disparage their religion and their fatherland, divest themselves of their traditions and beliefs, and to regard as sacred anything Western."

He founded the Muslim Brotherhood, out of which eventually sprang other fundamentalist movements like al Qaeda, Hezbollah, and Hamas.

Sunnis viewed the end of the caliphs in the 1920s with the same sense of revulsion that al-Banna experienced. Bin Laden made that clear in a video broadcast by Al-Jazeera TV shortly after the 9-11 attacks on the World Trade Center and the Pentagon.

"What America is tasting now is only a copy of what we have tasted," he raved. "Our Islamic nation has been tasting the same for more than eighty years, of humiliation and disgrace, its sons killed and their blood spilled, its sanctities desecrated."

Bin Laden and Zarqawi were especially offended by the ascendancy of Shiite dominancy in Iraq after Saddam Hussein's overthrow. Having been outside the circle of power for so many centuries, even though they were the majority by nearly four to one, the Shiites had developed a strict and independent hierarchy of their clergy. These already-existing, clerical-based social structures in Shiite communities were among the few organizations to survive Saddam's totalitarian regime. As a result, they more easily organized themselves politically after 2003. Under U.S. occupation, the Shiites were in, the Sunnis out.

As the Sunni-inspired insurgency grew into widespread terrorist violence, Shiites in power responded with aggressive counterterrorist operations against the Sunnis. The bodies of dozens of Sunni men with gunshot wounds to the backs of their heads showed up in the Baghdad morgue each week. Many of the corpses had their wrists bound with plastic police handcuffs. Although opposing foreign intervention in Iraq, Muqtada al-Sadr's Shiite Mahdi Army was attempting to build itself a strong presence in Baghdad's regular police force. The Shiite government of Iran actively supported Iraq's Shiite militias.

The long-festering bitterness between the two factions was apparent in a letter that U.S. intelligence intercepted between Zarqawi and members of the al Qaeda leadership.

"[Shiites] are the insurmountable obstacle," Zarqawi wrote, "the lurking snake, the crafty and malicious scorpion, the spying enemy and the penetrating venom. Shi'ism is the looming danger and the true challenge. They are the enemy. Beware of them. Fight them.

"If we succeed in dragging [the Shiites] into the arena of sectarian war, it will become possible to awaken the inattentive Sunnis as they feel the imminent danger and annihilating death at the hands of the Sabeans [a tribe regarded as nonbelievers]. Despite their weakness and fragmentation, the Sunnis are the sharpest blades, the most determined and the most loyal when they meet these Batinis [heretics] who are people of treachery and cowardice."

There was one other area of differences between Sunnis and Shiites in which they actually had more in common than not, that being the so-called "Twelver" phenomenon.

Shiites believed that the Twelfth Imam, who disappeared in A.D. 874, would return from the supernatural world to bring peace, prosperity, and a just and *global* caliphate to the world. The Twelfth Imam—who would return as the Mahdi, the Messiah—was comparable to the Christian Jesus. Sunnis believed the same thing, with one difference. They didn't think he was the old, in-hiding "Twelfth Imam" of the Shiites. Their Twelfth Imam had yet to emerge in history.

"Twelvers" were the more extreme members of each sect, primarily Shiite, who believed that the Twelfth Imam could be ushered back to earth only through fire and brimstone, by deliberate worldwide violence that would establish Muslim rule over the entire world, such as a nuclear war instigated by Iran. President Ahmadinejad was a committed Shiite Twelver.

I assumed the Twelfth Imam would present to each of the faithful his seventy-two virgins. I also thought that from A.D. 874 to 2005 was a long time for the Twelfth Imam to hide out. It seemed that the *National Enquirer* would have flushed him out by now.

My mind reeled. What I learned was disturbing. Could fanatical Islamists like these ever be suppressed? Why weren't moderate Muslims decrying the perversion of their religion? Why didn't the attrocities committed by Muslims against Muslims receive the attention that collateral damage caused by militaries of the west did? Terrorism and the onslaught of Islamic radicalism was becoming a pervasive and constant part of a world growing ever more dangerous with the proliferation of nuclear and biological weapons. How could it not one day ignite in flames and smoke— whether or not the Mahdi returned?

The American intelligence officer, a thin, bookish-looking captain with a crew cut, drove up to the entrance of the Italian POC and got out, dusting himself off from the drive over. He had called first to ask for a private confab with me. He spread a map on the hood of his vehicle and pointed to a sector of desert about twenty miles south of the airfield.

"We've received HUMINT [human intelligence] that insurgents have built a training camp in this region," he said. "Do you think your Italians can take a look-see for us with their Predator? The Global Hawk has been searching for two weeks and so far has come up dry."

No American Predators from Balad Air Base flew this sector because the Italians had it covered. Besides, things were so peaceful in the AO that the Dutch worked it on the ground with the Italian special forces—and neither did much patrolling way out in the desert. The war didn't set well politically in either Holland or Italy. Right after I hooked up with the Italians, Colonel Gentile confided in me that the war was so unpopular in Italy that soldiers were cautious about wearing their uniforms off-base. They were often harassed and sometimes confronted in the streets.

The fiasco during which insurgents drove the Italians out of Nasiriyah had caused a sharp drop in the popularity of the Italian government and their participation in the war. Since then, the single Italian casualty I knew of, the helicopter door gunner killed by the sniper, had caused so much panic back home that most Italian forces were afraid to do anything that might be considered hazardous or controversial. Their Predator wasn't even armed. While officers were eager, their government was risk-averse and was in the war only because the United States needed a "coalition." In many respects, the "coalition" in "coalition forces" was in name only; United States Marines, soldiers, and airmen did most of the fighting.

The captain from intel was hoping that I might intervene personally with Colonel Gentile. He was cautious about approaching the colonel directly because of a briefing I had previously delivered to his intel staff.

"The Predator crews are raring to go," I had begun in a dusty briefing room in Saddam's former HQ building at Ali. "They want to get at the bad guys—except bureaucracy has a chokehold on them. To begin with, they can't justify flying an entire mission for the sake of the coalition. The reason they bought the Predator system was to provide coverage for the Italian special forces that are supposed to be patrolling Nasiriyah and adjoining areas. As far as the Italian public knows, Nasiriyah is a peaceful city and their soldiers are sitting around watching the world go by and getting suntans. There'd be an uprising in Naples if it got out that their troops over here were conducting raids, arresting people, and occasionally shooting them."

The intel people shook their heads in frustration. "We're supposed to be allies," they complained. "The Italian special forces are so secretive they don't tell us shit about anything."

"It's not that they're secretive only with us," I continued. "The Italians don't even tell one another what they're up to. They don't want anything leaking out back home about what they may be doing. They're afraid of being crucified in the court of public opinion. That means they're sure not going to get involved in action that might be considered aggressive war fighting—at least not in anything that can't be kept to themselves."

It took an Italian general and six colonels at Ali Air Base to manage two Predators and their pilots and crews. That wasn't only because the Predator was the hottest thing going in the Italian air force and everyone wanted in on the action but also because, I suspected, all that brass was here to keep an eye on the program and make sure it stayed within certain boundaries.

Flying a Predator mission specifically upon U.S. request could be considered aggressive in Naples. However, since the elections, things had been so slow that Colonel Gentile and his pilots were chomping at the bit for something to do.

"We are dying of boredom, Major Martin," Lieutenant Orso protested. "We are supposed to be support—but our soldiers are doing nothing that requires support. We just fly around and around like a carousel."

Hunting for the insurgent training camp would be a *real* mission. I knew that Orso and Vinuti would be up for it. The intel captain refolded his map and looked at me.

"I think I can get Colonel Gentile to sneak it in on a routine recon flight," I said.

Damned if Lieutenant Orso didn't locate the camp that same night. It consisted of a half-dozen tents and crude shacks partly camouflaged in the middle of the desert. He relayed his video and the coordinates of the camp directly to American intelligence; it would have taken too long to go through channels and probably would have gotten lost in the process, or intercepted and cause a stir back in Italy.

The intel captain was elated. However, the quality of the infrared images shot through the Predator's early-generation targeting pod was grainy and poor. It would take a few more years before the Italians upgraded to the improved pod used on U.S. RPAs.

"I need more imagery to confirm this is a training camp and not a tribe of Bedouin passing through," the captain said. "It would help if they could shoot some daylight video before we call a strike."

Not many Bedouin parked SUVs around their tents or installed electric generators. Still, you never knew. I promised to see what I could do.

Colonel Gentile was hesitant about taking another run on the camp. Apparently, the general and six colonels had gotten wind of the operation and were afraid that if Americans bombed the camp based on Italian information, it might somehow end up hurting them.

"So . . . ?" the intel captain prompted.

I shrugged. "Grappa, two-hour lunches, and *domani*," I said apologetically.

I later learned that American F-18s bombed the camp. Orso grinned broadly when I told him the news; he had had a hand in contributing to the war rather than merely watching it. I knew how he felt.

Senior Airman Yob and I donned our helmets and flak jackets, tossed our bags into the cargo bay of the Italian HH-3F helicopter, and clambered aboard for the hop to Kuwait on the first leg of our return journey to the United States. Our job was finished here; the Italian Predator crews were "good to go."

Tears flooded my eyes as the chopper sprang into the air and morning sun spilled yellow veneer across the desert, which might have been almost lovely under un-war circumstances. They were not tears of sentimentality, to be sure: I was relieved to get the hell out of this Hades. They were caused by the spray of sand kicked up by the chopper.

Nonetheless, sentimentality or not, I would miss Colonel Gentile, Orso, and Riccardo Venuti and their aircrews. When I was "Matt the Cat" hauling hay on my dad's farm in Indiana and snapping up my head to gaze into the sky whenever an airplane flew over, I used to dream of flying all over the world and seeing everything there was to see. I remembered a phrase from somewhere, perhaps from Shakespeare, that ended something like "never this way come again." I doubted I would ever "this way come again," not to Ali Air Base anyhow, and I expected that my Italian friends standing on the taxiway waving and growing smaller and smaller would be the way I would always remember them.

I watched them and the runways gradually recede into the hazy distance as the helicopter poured on the coal and began to scream over barren desert at 120 knots, above the occasional oil well, goat herds, or burnt-out hulk of a tank. It was such an ancient land, this "cradle of civilization," supposed home of Adam and Eve and of Sargon, the Great King. I received a last glimpse of the ruins of Ur to one side of the base.

One of the last unofficial missions that Yob and I accomplished in Iraq was to explore the ruins. I had borrowed a truck in order to play tourist-for-a-day with Yob and a couple of other GIs. On the way, we stopped at the old Iraqi Air Force HeadBobbys. The main building was constructed of steel-reinforced concrete. Two holes had been drilled into the roof by a U.S. air strike during Operation Desert Fox in 1998. One of the bombs, a 2,000-pound "bunker buster," had penetrated all the way to the basement before exploding and gutting the place.

A nearby wall displayed a full-color, larger-than-life mural of Saddam Hussein and his glorious air force. It featured fighter planes swooping overhead, SAMs guarding the perimeter, radar and electronics detectors ever watchful for the enemy, and valiant ground forces alert and ready to stop any invader. Saddam, wearing his standard green fatigues and kitschy beret, watched over it all with a fatherly smile and his mighty arm outstretched. Above his head, as though fluttering in the breeze, were white doves of peace.

All evidence of Saddam's reign, I suspected, would be long gone within a generation, forgotten, melted back into the sands of time while nearby Ur still remained at least partly standing after nearly five millennium, the Ziggurat towering above it.

The ancient Sumer city of Ur thrived around 2600 B.C., dedicated to the moon god Nanna. According to *The Book of Jubilees*, the city was founded in 1688 *anno mundi* (year of the world) by "Ur, son of Kesed." In that same year, *The Book* added, wars began on Earth.

Ur was also mentioned three times in Genesis in the Old Testament. Many consider the city to be the birthplace of the patriarch Abraham.

Ancient history and archaeology told a more prosaic story. As the flood/drought cycles of the Tigris and Euphrates rivers became more dramatic, small farming communities of the Fertile Crescent began to consolidate into true cities where collective labor made large-scale crop irrigation possible. The first centralized governments and professional armies arose in places like Ur. A code of laws enacted under King Ur Nammu (2047–2030 B.C.) was one of the oldest such documents known to man, preceding the Code of Hammurabi by three hundred years.

From the Great Ziggurat fortress in a corner of the wall that surrounded the city, I could see the Royal Tombs of Ur. Excavations funded from 1922 to 1934 by the British Museum and the University of Pennsylvania uncovered a total of 1,850 burial sites, including sixteen "royal tombs" containing many valuable artifacts. Most of the loot ended up at the British Museum in

London and at the University of Pennsylvania's Museum of Archaeology and Anthropology.

Cuneiform (Sumerian writing) was carved into walls and tombs, along with modern graffiti in colored markers. "Habib loves Raisa and Sofia." Pottery debris and human bone fragments littered the grounds and composed parts of walls around the tombs.

Yob and I posed for photos with the other GIs. It reminded me of the scene in *Mars Attacks* where Martians pose for cheesy snapshots in front of Mount Rushmore while other Martians blow it up. Afterward, I climbed to the top of the Great Ziggurat fortress and looked down upon what may have been the earliest city in the world. I couldn't help considering how many other warriors over the centuries—Romans, Mongolians, Babylonians, Turks, British—must have also climbed this monument as visitors or as conquerors.

Temporary ones to be sure. Conquerors come and go. None remained at the top for very long. That seemed to be an unwritten rule of history, a realization that put a new perspective on America's presence in Iraq. I climbed down from the Ziggurat much humbled.

PART III

Balad Air Base, Iraq

CHAPTER 18

Back to the Sandbox

"THEY'RE COMING, PEOPLE!"

The base was being attacked and overrun. Enemy troops charged toward the POC and GCS trailers. Pilots, sensors, mission coordinators—everybody became a rifleman. We grabbed Kevlar helmets and M16A2 rifles and merged with security forces, maintainers, services, and anyone else available to form combat elements.

"Withdraw! Withdraw! Haul ass, people!"

Three fire teams made up my element. Each fire team consisted of four airmen and a senior officer. I was ranking officer for my team. I led it scurrying across the desert while the other two teams laid down covering fire. Retreating to a better defensive position. To my rear rose the fierce rattle of rifle fire, punctuated by yelling and the deeper, throatier barking of AK-47s as the enemy pursued.

A dry streambed opened up ahead. I hit the ground, rolling with my rifle tucked close to my body to prevent its flailing above shrubbery to expose my position. High ground gave way, and I plunged to the bottom of the shallow gulch and then sprang up immediately to look over the rim, weapon ready for action. The rest of the team spread out to my right or left. One of the airmen tapped off two quick rounds, his target unclear.

Seeing nothing to shoot at in the desert scrub, I held fire while the trailing team jumped to its feet and sprinted back to us, sailing over the gulley above our heads and leapfrogging on to the rear to set up the next cover in the bounding overwatch of withdrawal from contact. I suddenly spotted an enemy soldier charging out of the undergrowth into a clearing of low sparse grass, so near I could see the determined look on his face. He wore a black turban and combat boots. Our eyes met. He opened up with a burst of automatic 7.62mm from his AK.

The mujahidin filled my sights. I aimed for center of mass while his assault rifle continued to chatter and spit flame on full automatic. I squeezed my trigger three times, just as I had been trained. The guy stopped dead in his tracks, looked around as though confused, and then tumbled to the ground, apparently lifeless.

Back in Iraq on election day, when command feared a Tet-like offensive and an assault on the air base loomed as some possibility, I had considered what it might be like were I compelled to fight for my life in personal combat. This, I thought, must be it.

The shrill blast of a whistle stopped the action. A trainer wearing a black T-shirt and a black baseball cap appeared from the sidelines. "Knock it off!" he shouted, running back and forth blowing his whistle to catch the attention of all elements in the scenario.

The "enemy" I shot got up from the ground and dusted himself off. Others, both "friendlies" and "enemy," stood up out of ditches and brush to congregate around the instructor.

"That wasn't too bad," the trainer said. "Next time, however, the bounding team swerves to the left or right of the overwatch to prevent masking its fire. You don't want overwatch to have to fire through you. All right, let's try it one more time before we move on to the resource protection drill."

It was Warrior Week at Creech Air Force Base, a rare opportunity for the air force to practice ground combat skills. We used standard M16s loaded with blanks instead of live ammo, drove

Humvees, set up perimeters, and fought off guerrilla fighters. Just like real grunt soldiers.

"You're here because you've been selected for deployment to Iraq or Afghanistan," instructors stressed during orientation. "You may well need to employ the training you're enduring this week. It could save your life."

I was on my way back to Iraq—this time to Balad Air Base, near Baghdad, rather than to some isolated, relatively safe post like Ali. I had been flying Predator for about a year, been promoted to major, and now would command an LRE squadron in the heart of the insurgency. Launching and recovering aircraft for pilots flying missions out of Nellis was not what I preferred, but I had a way of fashioning duty to fit my expectations.

While Trish was not exactly comfortable with my leaving for war again so soon after having returned, she accepted it with her usual grace and with less stress than before. Over the next few years, she would get accustomed to my periodical dashes off to war zones. That was the nature of flying Predators.

Recent health issues had caused us both some concern. At my urging, she consulted a doctor about frequent leg joint pain that had plagued her for several months. We were both relieved when the doctor diagnosed her condition as a mild form of rheumatoid arthritis easily controlled with antiinflammatory medications. She was thinking about going back east to complete her postgraduate studies while I was away. That would keep her busy, and I wouldn't have to worry about her being stuck in Nevada away from most of our family members. War could really be tough on a soldier who left unfinished business back home.

In Iraq, I would need to concentrate fully on my duties. I doubted there would be any more two-hour lunches and certainly no more "*domani*" or "in its own time." I expected war this time to be spelled with a capital *W*.

Located about forty miles north of Baghdad, Balad Air Base was the largest and busiest in Iraq. F-16s and F-18s; C-130 and C-5 transports; more than two hundred Apache, Blackhawk, and Chinook helicopters; one Predator squadron; and some twenty-five thousand American troops called it home. In a typical month, as much cargo and five times as many people moved through Balad as through Dover Air Force Base, in Delaware.

Balad Air Base covered fifteen square miles protected inside a twelve-mile security perimeter. It was so large that it had its own "neighborhoods." Right in the middle of the Sunni Triangle, the most hostile region of Iraq, had sprung up a small American town. Four mess halls, movie theaters, a hospital where doctors performed four hundred surgeries a month on the wounded, gyms, a Subway sandwich shop, a Pizza Hut, a twenty-four-hour Burger King, two Post Exchanges stocked like Walmarts, and a miniature golf course that mimicked a battlefield with baby sandbags and little Jersey barriers, strands of concertina wire, and a tiny detainee cage at the end. Being stationed here wouldn't be exactly a hardship tour.

Miniature golf was about as close as a majority of the troops on base would ever get to a battlefield. Few would ever interact with the Iraqis; some would never even see an Iraqi. Nonetheless, insurgents fired several mortar rounds a day at the base, causing an injury now and then and a few fatalities over a period of two years or so, prompting Balad's nickname of "Mortaritaville." Still, the odds against an individual getting hit by a mortar round were about the same as being struck by lightning somewhere inside the "Tornado Alley" of Oklahoma and Texas.

It was a different matter outside the wire, however. Violence surrounded Balad on all sides. Within past months alone, suicide bombers, IEDs, car bombers, snipers, and ambushes had taken a terrible toll on civilians and military alike. Shiite cleric Muqtada al-Sadr had led tens of thousands of marchers through Baghdad protesting the U.S. occupation. On the outskirts of the march, insurgents ambushed and killed fifteen Iraqi soldiers traveling in a convoy.

Nineteen Iraqi soldiers were executed in cold blood at a football stadium; another nineteen were slaughtered in Haditha in the same manner. Car bombs at an Iraqi Army recruitment center, at a Shiite mosque, and at a market in Shula and another in Baghdad left 114 dead and 297 wounded, most of them civilians and many of them women and children. Suicide martyrs also got in their licks by killing 140 people and wounding that many more. One suicide attack against a Kurdish funeral procession killed and wounded 55 people.

Since the beginning of April, insurgents had killed 10 U.S. soldiers, 57 Iraqi soldiers, and 451 civilians and wounded more than 500. Obviously, civilians were getting the worst of it. Internecine warfare at its goriest.

Two years ago when I was flying navigator in RC-135s out of Crete at the beginning of Operation Iraqi Freedom, the U.S. Air Force had had a single Predator flying daily over Iraq. Predators and RC-135s were looked upon as "high-value airborne assets" when it came to reconnaissance, but we were almost as limited in 2003 at providing up-to-date information on targets as we had been during Operation Desert Storm in 1991. Captain Jon Hodge, now a Predator pilot, recalled 1991 when he had been a B-52 bomber pilot.

"After six hours of flight en route to a target," he said, "we had next to nothing in the way of additional intelligence added to the pre-flight briefing. In flight, we were lucky to have voice updates from high value airborne assets."

By the time Capt. Roger Brent and I hopped "Catfish Air," a C-130 that made regular service from Kuwait to Balad Air Base in May 2005, "high-value airborne assets" had increased to four Predator combat air patrols (CAPs). Four RPAs in the sky at all times, each on a twenty-hour sortie feeding continuous live video on targets and situations to commanders and cockpits throughout every phase of a mission. Battle damage assessment could be relayed so quickly that the original strike aircraft could reattack almost immediately if necessary. Predators were our "unblinking eye" in the sky.

I was being assigned as commander of the 46th Expeditionary Reconnaissance Squadron. Predators. The squadron consisted of four pilots: Capt. Roger Brent, Capt. Joe Henson, Capt. Bobby Rangler, and myself. Support included four sensor operators, one medical technician, a dozen supporting contractors, and about sixty-five maintainers, who kept the planes serviced and running. As LRE, my pilots flew the airplanes only during takeoffs and landings. Once we had a plane safely in the air and configured for combat, we turned over control to a combat pilot 7,500 miles away at Nellis. Then we pretty much sat back and waited until the plane returned and we landed it.

Not much excitement in that. Even before I reached Balad, I was starting to feel the same impotence that our LRE must have felt when I was among those back in Nevada actually flying missions, getting the glory.

So-called "Catfish Air," into which Captain Brent and I were strapped, circled the gigantic airfield with its twin runways for fifteen minutes while we waited for clearance to land. I watched out the window as a Blackhawk landed and an unmanned Predator took off, two Hellfire missiles underneath its wings. An awesome sight.

Next in the pattern came an army RC-12 Guardrail bristling with sensor antenna. In quick succession followed an F-16 fighter, a C-130 Hercules, a C-17 cargo jet, and a Russian 1L-76 freighter with its bulging glass nose.

"It's like putting Chicago-O'Hare right in the middle of Iraq" was how one airman described it.

It reminded another of the Atlanta airport. "Except in Atlanta, there's no one shooting at you."

I was likewise astonished at the size and complexity of the base. Since leaving Crete, I had been moving incrementally nearer the war. From flying RPAs out of Nellis, which put me mentally and emotionally if not physically in combat, to relatively peaceful Ali Air Base with the Italians. And, now, on the ground in Balad, the most active combat site in-country.

"All in all," Brent commented, gazing pensively out the window, "I'm ready to go home. This place is a shit hole. Matt, I got a wife and five kids. A good Catholic boy like me belongs here about like a mujahidin in Tel Aviv."

That gloomy statement, although I didn't know it yet, would prove to be a prognostication of things to come in Brent's adaptation to the war zone. He should have thought about the demands of an air force career long before now.

Physically, Brent was a larger man than I, with a soft look of face and body. He was about forty years old with short-cropped brown hair flecked with gray. To me, he looked like a pharmacist or a small-town grocer. He was right, he didn't belong here. But we still had a job to do.

Brent's mention of family made me think of Trish and how much I already missed her. A serviceman's wife puts up with a lot of hardship, especially during wartime. I knew it must be easier in many respects to go off to war than to wait at home for a loved one. Trish's diagnosis of rheumatoid arthritis made our separation all the more poignant. Although doctors assured us that the condition could be controlled, she was still often in pain when I left.

"I'll be all right, Matt," she tried to comfort me. "You just worry about taking care of yourself."

"Honey, promise me you'll go back to the doctor if it doesn't get better."

"I will, Matt."

I made a mental note to call her tonight after we landed.

Catfish Airlines finally entered the pattern, landed, and taxied up to the ramp in front of Operations. Brent had the look of a homeless hound on his face. I didn't think his morale was likely to improve, starting out like this.

The entire 46th Expeditionary Reconnaissance Squadron was waiting to welcome their new commander to Balad. A skinny Oklahoman with a baggy flight suit hanging on his thin frame stepped forward wearing a big "aw shucks" grin. Captain Bobby

Rangler, as I soon found out, was never happier than when he was in the pilot's seat chasing down and killing bad guys.

"Welcome to Mortaritaville, sir," he proclaimed with a formal salute. "Sir, we are going to kick some serious ass."

For all Capt. Bobby Rangler's enthusiasm, "kicking serious ass" was not a realistic option. At least not yet. Since our planes could take off and land only with line of sight due to the delay in satellite transmissions, thus the need for an in-country LRE, the squadron's job was simply to launch and recover aircraft and remain on standby. Pilots at Nellis did all the serious flying.

Until shortly before my arrival at Balad, the LRE kept a plane in the air over the base at all times as quick reaction against mortar and rocket attacks. It worked for a time, until the bad guys wised up and started using timers to fire off rounds by remote control. After that, the local Predator might spend weeks and weeks flying with nothing to show for it. Finally, the CAOC at Qatar decided that the mission was a distraction and the LRE should focus on its primary task. Whether I could revive the practice from its quiet death remained to be seen.

Each of my pilots expressed reaction to the restriction according to his temperament.

"Sucks" was Bobby's backlash. "A bull without gonads."

"We need to get back into the fight," said Capt. Joe Henson.

Henson's call sign was Statler, not only because of his last name but also because he resembled a younger version of the Jim Henson muppet character Statler. The captain was a few years younger than I and already starting to bald. So he shaved everything except for some fuzz above his ears. He was a sharp guy, although he walked around with a faraway look that said his tour in Iraq had started to wear on him.

Captain Roger Brent was solidly predictable in his pessimism. "Get this shit over with and get back to civilization."

And me? As always, I tried to be completely level and responsible. Whatever my assignment, I intended to do it properly and honorably and to the best of my abilities. If along the way I just happened to stretch my wings to encompass more territory, well... I soon became convinced that the pilots in-country could be more responsive to tactical demands than the pilots who were thousands of miles away. We had a more personal involvement when we shared the risk. The primary drawback was that line-of-sight operations were restricted to a radius of a hundred miles, since satellite control could not be utilized, while pilots at Nellis could roam the entire country.

In my opinion, line of sight and remote satellite each had its advantages and disadvantages and should both be utilized within their parameters.

"I think we can fudge a couple of hours at the beginning and end of each flight to run local missions," I reassured Bobby.

Without "kicking ass," LRE operations quickly settled into a routine, disrupted occasionally by mortar attacks and the odd in-flight emergency. The previous LRE team had crashed two airplanes during its tenure, at a cost of over eight million bucks to American taxpayers. I intended to do better.

CHAPTER 19

Johnny Rico

A S SQUADRON COMMANDER, I lived in my own private trailer in a little "trailer park" and was assigned a Nissan Pathfinder for commuting to work. Air force regulations mandated a minimum of twelve hours' crew rest between flights. On a typical day, however, I was up by 0300 hours. Following a workout at the gym and breakfast at the chow hall, I reported for work before 0600 hours, in time either to catch or launch a flight.

After meetings with either Brigadier General Hanson, commander of the 332nd Air Expeditionary Wing, or with Colonel Booth, who commanded the 332nd Expeditionary Operations Group and was my immediate supervisor, I had a late lunch and was in and around the ops center until Statler Henson or Bobby kicked me out at 2000 hours (8 p.m.).

"Let's get out of here," Bobby would drawl in his best Okie twang, hitching up his baggy flight suit. "We don't want you violating regulations and your own rules. Statler can handle it."

As ranking officer below me, Captain Henson was the squadron's director of operations (DO). He was also, by preference, generally in charge of the night shift.

"Major Martin, I'm checking on you," Senior Airman Kimberly Joan often teased. She was my sensor operator when I flew, an attractive twenty-two-year-old black woman who had taken it upon herself to be my mother hen. Like Bobby, she was concerned that I was working too many hours and ignoring regulations. "I'll take care of you, Major. You just listen to Airman Joan."

Predators were temperamental aircraft that required constant attention, both mechanically and in regard to weather and the elements. With four aircraft in the air at a time, each on a twenty-hour sortie, I sometimes had to jump through hoops just to keep everything sorted out. I might be running a functional check flight on one aircraft with Kimberly when Bobby or Brent, whoever was on duty with me that day, would stick his head into my GCS to announce that an airplane was returning with a sticking fuel feed solenoid. If it got any worse, the plane might not be controllable and would crash. That meant shutting down one of the runways for an emergency landing.

On some other day, a big sandstorm might swirl across the flats toward the air base like an ugly brown cloud from America's Dust Bowl days. Whenever that happened, whichever Nellis pilots were flying turned their planes over to us like hot potatoes. Visibility would be about a Bobby-mile through the infrared camera or about zero with the naked eye. During one such storm, Captain Brent's concentration slipped and he bounced the Predator on the runway, scraping the tailboards. That meant redlining the plane for a few days and flying short until it could be repaired.

"I'm just not with it, Matt. Sorry," Brent apologized in that hangdog manner of his. "I keep thinking of my wife back home trying to hold the family together while I'm gone."

I expected him to quit one day, just get up and walk away and to hell with the consequences. I often wondered how he had completed Predator school. How he had become a military pilot to begin with, an attainment that required focus and discipline. I spent an hour or so nearly every day just trying to keep him centered. Sometimes that wasn't easy, as I had to keep a desk or a table between me and the awful blast of his kimchi breath.

Kimchi was a form of fermented cabbage common in Korea and other parts of Asia. Somehow, Brent had acquired a fondness for it; I knew his wife wasn't Korean. He snacked on the stuff for lunch nearly every day. Some days he reeked so badly that it was hard to stay in the same GCS with him. I pitied his poor sensor operator.

"Roger, we all have families at home," I reasoned. "My Trish is facing a medical problem all by herself. But we have a job to do here."

"I hate this lousy place. I should have my head examined for jumping into this shit."

"But you did, Roger. That's the reality. So do your job and don't let the team down."

Each time he promised to cowboy up—but then it was the same thing all over again the next day.

By the time my shift ended and I turned over the operation to Bobby or Statler, I was so tired that I was almost falling asleep. I would stop at the Post Exchange on my way to the trailer to call home and check on Trish, then have late chow and work on a War College correspondence course. I often awoke with my head down on my desk.

My old Nissan had gotten pretty well beat up by previous drivers. One morning it refused to start. I slapped the wheel in frustration and radioed my DO, Statler Henson, for a lift. He showed up looking more Statler than usual.

"There was a Predator crash last night," he said.

My first thought was Brent, but I didn't want to indict him offhand. "Who was flying?"

"Me."

My speculation is that Statler had committed a classic error partly as a result of the poorly designed off-the-shelf interface that was the Predator cockpit. He landed his aircraft and taxied clear of the runway. As he initiated the procedure to turn on the engine cooling fan to stabilize oil temperature prior to shutting down the engine, he inadvertently typed in an incorrect sequence on his keyboard. M0-M1-M2 instead of M1-M2-M3. One incorrect key to the left.

Instead of turning on the engine cooling fan, the aircraft responded to the incorrect stimulus and engaged its autopilot. The next thing Statler knew, the Predator was attempting to fly to the next waypoint on its programmed route—even though it was on the ground.

It applied full power, ran off the edge of the taxiway, bounced into the air a few times, and ended up nose down in a ditch. The crew chief ran up and hit the engine kill switch on the side of the airplane. Repairs would cost over a million dollars. So far, with barely a month on the ground, I was well on my way to crashing more airplanes than the previous LRE commander.

Normally, a pilot who crashed an airplane would be suspended pending an investigation. I didn't have time or the manpower for that, not with only four pilots, including myself. One of my GCSs would have to go unmanned for twelve hours a day, meaning that sorties would be cancelled.

I spent the day and into the night explaining to General Hanson exactly what happened, why it was such an easy mistake to make, what procedures and remedial training I would enact to ensure that it didn't happen again, and why Captain Henson was an indispensable part of my team.

The general bought off on it. Statler let out a sigh of relief.

When the replacement aircraft came in, I dubbed her *Ruby* and sent Trish a picture of me in front of her with Trish's nickname on the fuselage. She was delighted.

Captain Bobby Rangler the Okie had come up with the idea. Johnny Rico in the movie *Starship Troopers* always wanted to kill "bugs," giant insect-like invaders. My 46th Expeditionary Reconnaissance Squadron therefore became known as "the Johnny Rico Squadron." Bobby, I assumed, saw himself as the hero.

So far, however, we hadn't hammered any bugs. That wasn't our job. Instead, we had to sit back and do nothing except launch and recover while enemy mortars beat us up day and night. We were suffering far more attacks than usual, reflecting the violence in Baghdad, Sadr City, and elsewhere. One of the more severe shellings occurred during a practice change-of-command ceremony on the parade field. Eight coordinated rounds screamed in. Four of them impacted in the housing area that adjoined the parade field and wounded four airmen. The change-of-command drill was abandoned while airmen ran in all directions for shelter, a truly surreal experience.

Different patterns of wailing sirens announced the various codes signaling a mortar attack. A long, wavering tone like that at the beginning of *On the Beach*, the old end-of-the-world movie, announced a round launched—*Alarm Red*. It chilled the blood, no matter how many times you heard it. You were supposed to dive into the nearest reinforced bunker and cower there until *Alarm Yellow* sounded—a long, straight bleat. *All clear.*

The alarms were connected to the army fire control network. As soon as a mortar round became inbound, the fire control radar system that ringed the base detected it and within seconds determined where it would land. *Alarm Red* went off, followed by a computerized loudspeaker voice that warned residents of that sector to take cover. When you heard that eerie mechanical voice saying "Incoming! Incoming!" you needed to hit the deck because you were under the crosshair, and the trigger had already been pulled.

One of my maintainers, Master Sergeant Adams, was getting a haircut in the barbershop trailer next to the Post Exchange when a round landed less than twenty feet away. It shattered glass and knocked him out of the chair. He got up, brushed himself off, and, at a hard run, followed the barber to the nearest bunker until the *All Clear* sounded.

"It was a closer shave than I bargained on," he remarked later.

Night attacks were the most harrowing. No one slept soundly after we had a few nights of explosions. The ear-shattering chill of *Alarm Red* jarred you awake. "Incoming! Incoming!" And you held your breath, not knowing where the next shell might land. Maybe in bed with you.

At 0300 hours one morning, another of my maintainers who had just gone off-duty was walking from the gym to his room when a round landed on the gravel pathway ten feet away. The explosion knocked him off his feet and sprinkled his side and belly with shrapnel. An ambulance hauled him away. By that afternoon, he was being med-evac'd to Germany. He was conscious when he left Balad and the prognosis was good that he would recover. He was the first casualty on my watch.

"Sergeant, I want to see you back here in a few months," I scolded at his bedside before he left, only half-joking.

"I'll be back, Major. You guys need me."

With spirit like that, how could we lose? Still, the United States had 150,000 troops in Iraq, and some were going home in body bags nearly every day or shipped out maimed and disfigured. All of us were negligent if we failed to provide the best support possible to protect them. I, especially, continued to feel that we were not utilizing Predators to their fullest capacity. There was so much more we could be doing, *should* be doing.

Bobby was furious.

"Matt, Johnny Rico needs to kill some bugs."

Most of the squadron felt the same way. The attitude often spilled over into comments that I happened to overhear or were passed on to me.

"We could declare victory and go home if we bombed this shit hole back to the Stone Age."

"Give me a supply of Hellfires and free rein, we could stop this crap within a month."

Nothing illustrated better the need for line-of-sight operations in Iraq than the incident of the fifty-five-gallon oil drums. I had promised Bobby at the beginning that we would fudge hours off flights dedicated to the POC at Nellis in order to fly our own local missions. During aircraft recovery, I began to steal an hour here and there to take a look-see around Balad's perimeter if an airplane had a little extra fuel left over. It didn't take a lot of spare gas at a burn rate of three gallons per hour.

During one of these purloined flights, I came across a couple of oil drums cast into a grove of palms within easy mortar range of the base. They were slanted toward the airfield in such a way that made me think they might contain mortar tubes. If so, all insurgents had to do was sneak out in the middle of the night, lift the covers off the barrels, fire a round or two, re-cap the barrels, and disappear.

I sent video to the army intel guys on the other side of the base and waited. Nothing happened.

A few nights later, Balad sounded *Alarm Red*—base under attack. I heard a couple of explosions go off between the two runways, the sounds reverberating and echoing across the air base like claps of thunder. The army fire control network pinpointed the fire as originating from the vicinity of the oil drums. The POC at Nellis was flying a Predator armed with Hellfires in the area. However, by the time I put the pilot at Nellis in communication with the JTAC at Balad, everything had gone *All Quiet on the Western Front*. The insurgents who had uncorked the drums and fired at the air base were long gone.

Frustrating. Damned frustrating. I knew we could catch those guys if we could just rededicate a Predator to perimeter security. Instead, we continued to live on the edge of our nerves.

Whenever intelligence received information of a possible pending attack, the base commander declared Uniform Condition 3. Not quite an *Alarm Red*, but preceding *Alarm Red*. Everyone on post was required to wear a helmet and flak vest under this condition. Our GCSs were located inside hardened aircraft shelters, which meant we didn't have to wear combat gear while flying. However, we donned it to and from work.

After my shift ended and Bobby took over for the night as DO, I drove to my trailer wearing a flak vest in the stifling heat of the Iraqi summer while hoping the air conditioning in my Bobbys wasn't on the blink again. I was sleeping soundly with functioning air when a deafening explosion all but jarred me out of bed.

Nine more blasts followed, each sounding closer and more threatening than the previous. Windows rattled, papers fluttered off my desk, and stuff vibrated off shelves. By the third detonation, I was ready for combat, breaking all previous records for getting garbed out. I sprawled on the floor with my head buried underneath my arms, expecting the next shell to crash through the roof.

Then it occurred to me that I hadn't heard any sirens or the scary computer voice. I crawled over to the radiotelephone. Bobby answered my call almost immediately. The strained tone of my voice prompted a burst of laughter.

"Why, shucks, Matt. Don't you remember? Them are controlled detonations. Fireworks. Happy Fourth of July."

"Oh."

I was still so wired that I probably wouldn't be able to get back to sleep until at least Thanksgiving. I pulled up a chair at my desk, rearranged the mess caused by the explosions, and began work on a concept of operations (CONOPS) about how we could better utilize Predators to protect the air base. If those fireworks *had* been salutations from the enemy, an eye in the sky controlled by my personnel on-site would have promptly located the source and extinguished it.

It was nearly dawn by the time I quit. I yawned and crawled back into bed for another hour, feeling sheepish about wearing my flak vest but wearing it anyhow. Just in case.

Czar Nicholas of Russia had thought his army too powerful to fail when he launched the Crimean War. His arrogance and rigidity, his refusal to change tactics or strategy when it was clearly demanded, ended up in his defeat and his army being pushed all the way back to Russia. I began wondering if perhaps the United States was too rigid to do counterinsurgency. Maybe we shouldn't be in this kind of war business if we couldn't change, be more flexible.

General Hanson rotated out as commander of the 332nd

Wing, replaced by Brig. Gen. Frank "Gork" Gorenc. A gregarious bear of a man, Gorenc was six and a half feet tall, had shoulders so wide they touched both sides of a doorway, spoke with a Slovenian accent, and would get right up in your face and throw an arm around your shoulders to better score a point. He had spent his career rising up through the ranks of the "fighter mafia," that group of influential fighter pilots who had wrested control of the air force from the bomber pilots who ran things during the Cold War. Unlike many other fighter pilots, however, Gorenc embraced the unmanned aircraft mission.

Prior to deploying to Iraq to take over as 332nd Air Expeditionary Wing commander, Gorenc became the first general in the air force to receive a checkout in the Predator. The senior officer course was only a few days in length, which meant he was not qualified to fly solo combat missions, a limitation he intended to remedy at Balad by getting in as much Predator flying as he could. That immediately made me, the guy who could get him on the flight schedule, his favorite person on base.

I liked Gorenc at once. If anyone would understand what I was trying to accomplish with my base defense plan, it was him.

CHAPTER 20

Wish You Were Here

BACK AT NELLIS, THE GCSs had been in constant use, but here they were used only about twelve hours a day between the two. I knew we had the capacity to add a local sortie and still be able to launch and recover on the LRE schedule. In order to prove that we could do both, I made sure that our aircraft were always ready to go and no sorties ever failed due to maintenance problems. My maintainers kept a "hot spare" ready to go; I drilled them in a new procedure that allowed us to quickly switch to the spare aircraft if the primary developed a glitch. I drove myself and expected pilots at the POC in Nevada to keep up. Rumor had it that I was becoming very unpopular back at Nellis because of the high operational temp I demanded.

"All these mortar attacks against us have a pattern," I explained to General Gorenc. "The insurgents already have the mortar base plates planted out there. They sneak up at night, fire off a round or two, and then disappear. Even if we locate them within a minute, we're not integrated enough between Nellis and our local operation to get eyes on, establish PID [positive identification of hostile activity], and take action. What we need is an integrated base defense plan."

"The CAOC is not going for it," General Gorenc responded, leaning in to me. "At least not yet. Matt, I'm all for it. When the time comes—and it will—I want you to be ready."

"I'll have a plan, sir. I won't give up."

The general gave me a broad grin. "Major Martin, I would not expect *you* to give up on anything."

I was out to my usual late lunch one afternoon when *Alarm Red* went off. I sought cover and waited until *Alarm Yellow* sounded, at which time mission essential people could report back to their duty stations. I jumped into my old Nissan, which fortunately kicked over, and headed for the Predator hangar in our hardened aircraft shelter. I arrived just in time for another *Alarm Red*. A mortar shell erupted in a column of black smoke and soil next to the nearest runway. People scattered for cover.

Captain Roger Brent, who was in the process of recovering a Predator, was burning up spare fuel by patrolling the perimeter. He immediately diverted the plane to the north side of the base and the calculated point of origin (POO) of the mortar fire. He arrived just in time to spot a white pickup truck speeding away. He gave chase while coordinating with an army quick reaction force (QRF).

Following Predator's directions, the QRF quickly cornered the pickup and dragged two suspects out of it. I could have kissed Brent, kimchi breath or not. This was just what I needed to demonstrate how my plan could work.

"We'd do more of this if we went to a line-of-sight mission," I appealed to General Gorenc.

"Work up a concept of operations around this example," he directed. "I'll shoot it on up the chain of command to the CAOC."

"I've already been working on it, sir."

As far as weapons systems went, RPAs—and in fact all unmanned systems—were new arrivals in the game. When Predator was being developed and refined in the 1990s, certain aspects of the program were kept hush-hush. Top secret. The U.S. government didn't want potentially hostile nations knowing about our capability to spy on them. A decade later, Predator suddenly sprang out of the shadows onto the national stage and into the headlines. It seemed that everyone from the newest freshman congressman to the reporter from the Podunk *Picayune* wanted a piece of it. As commander of an expeditionary reconnaissance squadron in Iraq, I reluctantly found myself at the center of attention.

An air force officer, especially one in a high-profile business such as RPAs, wore many hats. It was almost enough to make you schizophrenic, if you weren't already. Not only did I have a squadron to run, I was also expected to attend dinners with generals, entertain visiting politicians and VIPs, and now, it seemed, make myself available to the media.

In recent years, print and electronic media alike had lost some of the prestige they had enjoyed when World War II reporters like Ernie Pyle were writing carefully worded dispatches without the blood and gore that TV later depicted. During the Vietnam War, reporters were encouraged to take the war home to the public on TV evening news, complete with heart-rending shots of American boys being maimed and killed. All the major cable channels and networks showed Somalis dragging mutilated bodies of U.S. Rangers through the streets of Mogadishu. Reality TV programs focused on the brutality and chaos that was the war in Iraq. All this tended to make most soldiers wary and suspicious of media motivation. News about the war appeared to be more about political controversy at home rather than whether or not national interests were being advanced or harmed.

I was naturally hesitant when a camera crew from Fox TV showed up at Balad to shoot a "package" on Predator operations. General Gorenc offered encouragement.

"Matt, you will give Fox the splendid dog and pony show," he exulted with his big hearty laugh.

The whole thing took about an hour to film. The producer selected the slant, wrote the script, and staged it. The Fox camera crew decided that I was the new Tom Cruise "Top Gun" of the virtual battlefield when the producers discovered that I was one of only a handful of senior Predator pilots who served as both mission commander and LRE squadron commander. I appeared wearing my standard Randolph Aviator shades, looking professional and experienced while delivering the standard unclassified briefing. We toured Predator facilities and I showed the camera declassified video of a Hellfire shot. Senior Airman Kimberly Joan and I manned a GCS and pretended to take off an aircraft.

Fox left. We all returned to business as usual and thought nothing else about it. Just part of the job in the modern world, feeding the twenty-four-hour newscasts and their insatiable appetites for novelty.

A couple of weeks later, I received in the mail a three-by four-foot poster-board photo montage from a viewer of the Fox news special. Apparently from a chamber of commerce banquet, the photos portrayed a bunch of rich people dressed up like their favorite celebrities guzzling champagne and having a high old time. I assumed I was intended to hang the montage in the HAS for my squadron's pleasure and edification. All the revelers at the banquet signed the poster, as though wishing us a good, happy war. "Thank you for being over there."

Why would anyone with an ounce of common sense assume that something like this would make my people content in being at war?

In return, the viewer who sent the poster requested that my squadron pose with it. Air Force Public Affairs was all too happy to oblige. A photographer came down, lined us up around the poster, and took a few snapshots. "Wish you were here."

The back side of the poster turned out to be more presentable. It featured a large pen-and-ink sketch of an air force honor guard. That was the side that ended up on display in the HAS.

Later that same day, the reality of war returned, along with a new realization of the restrictions the squadron faced in dealing with base security. We were stealing a few minutes from the front end of one of our launches for Nellis when the JTAC, call sign *Darknight 26*, requested that we look for a taxi that was following a U.S. Army convoy through the town of

Balad. I climbed the airplane to altitude and soon located the taxi. While I was directing a QRF in Humvees to intercept the taxi, the mission commander at Nellis demanded that I give up the chase and turn the flight over to one of his crews at the POC. I had no choice but to comply and let the cab get away.

My frustration at the system didn't end there. Germany experienced a satellite breakdown. Four of my airplanes automatically went to "lost link" and, as programmed, flew themselves home. We captured two, landed one, and were about to land another when the satellite link came back on-line. Three of the airplanes were able to resume their missions, although at a loss of several on-target hours. Just another reason why the CAOC should accept my concept of operations when I completed it and sent it up the chain of command. Line of sight would have kept us in business during the breakdown.

The next day, problem child Capt. Roger Brent landed a drone without tower clearance. Pure carelessness. Tower filed a complaint. With everything going on, perhaps I should have annotated the photo of my squadron to the chamber of commerce people with "Look how much fun we're having."

The shot had been perfect—until the last five seconds.

As usual, my crews at Balad were standing by in their GCSs, monitoring flights for Nellis POC pilots and waiting to take over to land them. I had just walked in to check on things when a platoon of marines burst onto the command radio net with excited complaints about taking mortar and RPG fire near Al Quim, on the Syrian border. The mission commander in Nevada diverted an RPA to the scene. Even if we had had line of sight, we couldn't have helped; the marines would have been out of our range.

The Predator's cameras relayed to our screens a vast expanse of desert broken by dry wadis and patches of green along an intermittent stream. The marines had pulled into a defensive perimeter on a dirt road that crossed the creek. As we watched, a mortar shell exploded a geyser of sand outside the ring of Humvees.

Marines in the turrets of their vehicles returned fire with mounted machine guns, spraying a low hill south of their position with bursts of red tracers.

By the time our Predator caught view of the insurgents beyond the hill, they were tossing their tubes onto the flatbed of a bongo truck and clambering on after them. A marine Cobra helicopter that had also been scrambled decided it would take the shot. The ground was open terrain with no risk of collateral damage.

It was an awesome spectacle, even on our screens. The chopper thumped into sight. Insurgents spotted it and began to leap off the truck and scatter for safety. The truck disappeared in a roil of dust and smoke as the Cobra rolled in hot, rockets and 20mm guns blazing.

When the picture cleared on our monitors, we saw the abandoned truck huffing smoke, its windows busted out and bullet holes pocking the cab. Not all the bad guys had escaped. The driver's body hung bloody out his open window. A couple of other bodies resembling twisted bags of rags lay sprawled on the ground.

The Cobra made a second pass, chewing up the countryside where the fleeing insurgents were hiding. As soon as it swept on, a second truck that had somehow escaped scrutiny roared out of concealment. Desperate insurgents piled onto its flatbed from every Bobby. The truck and its merry band of bad guys raced toward the nearest town about a mile away, fishtailing and kicking up a funnel of dust. By the time the Cobra caught up with it, the truck was already safely inside the town and parked in a suburban neighborhood. As usual, the guerrillas thought they could avoid retaliation by mixing with civilians where we dared not fire at them.

"That takes some sack," Bobby commented admiringly.

"They do indeed have sack," I agreed.

The insurgents loitered around the truck, obviously figuring we couldn't touch them. They, however, had no idea they were being watched by an invisible eye in the sky. One that was armed.

The Cobra pilot recognized the situation and returned to base. A brief chatroom discussion ensued between the Predator pilot and the marine battalion being supported. The pilot suggested the target was feasible because the vehicle was parked across a large courtyard from the nearest house, which put the house out of range of the missile's fragmentation pattern, and pedestrians appeared to be staying clear of the vicinity of the truck.

The marine ground commander through the JTAC, who was also monitoring the Predator's view of the target, gave clearance to take the shot. The Predator crew spun up the plane's Hellfires, set up an attack run, and fired a Special K, sending the missile on its way with a twenty-three-second time of flight.

Everything looked perfect until . . .

Two kids on a bicycle unexpectedly appeared on the screen approaching the truck and the insurgents. Both were boys. One appeared to be about ten or eleven, the other—possibly a younger brother—was balanced on the handlebars. Tooling along on a summer day laughing and talking.

"Oh, God! Not again!" escaped my lips. Two separate images filled my mind simultaneously.

The first was both a picture and a feeling of peddling my little sister on a bicycle like that on a summer day long ago in Indiana. I even felt the sweat on my face as we tackled the hill near our house. I smelled little Trish's hair, heard her laughter all over again. My sister was also a Trish, part of the reason for my wife's nickname, Ruby. Like the Iraqi kids, we had had no idea of impending danger.

The second image was of my having maybe killed the old man in front of the wall while taking out the Rocket Man. That day had plagued me ever since. And, now, with the kids, it was like déjà vu, only ten times worse. And there was nothing I could do to save them, nothing anyone could do. Mesmerized by approaching calamity, we could only stare in abject horror as the silent missile bore down upon them out of the sky. It could not be diverted without the risk of causing even greater carnage.

I caught my breath as the missile impacted and the screen pixilated. I heard a muted scream from Kimberly. Brent shrieked, "No!" He had kids about the same age.

When the screens cleared, I saw the bicycle blown twenty feet away. One of its tires was still spinning. The truck was a mangled scrap pile of wreckage. The bodies of the two little boys lay bent and broken among the bodies of the insurgents.

The responsibility for the shot could be spread among a number of people in the chain—pilot, sensor, JTAC, ground commander. That meant no single one of us could be held to blame. Still, each of us shared in the tragedy.

From the gloom that descended upon the Balad LRE, an uninvolved observer might have concluded that our collective firstborns had just been taken out back and shot. Pilots and sensors congregated in solemn denial around the GCS screens, still in shock over what we had just witnessed—another of the dirty little horrors of war that lost none of its impact whether you were actually there or you viewed it all by remote. Death observed was still *death*.

Bobby the Okie, with his ability to philosophically detach himself, was the first to find words. "Break up the pity party," he scolded. "What's done is done. No good can come from obsessing about it. It'll only distract us from doing our job."

Were it that easy. Senior Airman Falisha Rexford, Bobby's sensor, turned her head to one side as tears filled her eyes. Tiny, dark haired, and eager, she was as ambitious and occupied with her job as Bobby, but the incident had caught her off-guard and exposed her vulnerability.

"We're as bad as terrorists if we laugh off the deaths of those kids," she snapped back at what she apparently assumed to be her pilot's insensitivity.

Bobby turned on her, angrily hitching up his flight suit. "Do I look like I'm laughing, girl? The only thing I said was that we can't obsess over it."

"Look," Falisha said more calmly, "I won't hesitate to do my job when the time comes. But a thing like this . . ."

Bobby sighed, working on his patience. "What happened was unavoidable. Bawling around like baby calves that have lost their mamas is not going to bring 'em back."

Captain Brent suggested we all go out and throw one hellacious drunk—except we couldn't even do that. No booze was allowed on base.

"It's enough to make us psychotic," he whined, sounding on the verge. I half expected him to plunge over the edge any day now.

I recalled one of those endless mind-numbing meetings that I was required to attend at which an F-16 squadron commander uttered one of the most profound statements I had yet heard about the war.

"Every time we drop a bomb," he said, "it's a strategic failure even though it may be a tactical success."

Killing the kids was like that. The tactical mission succeeded, but the strategic picture had been blurred. The world press would use it as propaganda against us. We would even use it against ourselves.

"What happened could not have been prevented," Bobby rationalized. "Things like this happen in war. We used every precaution."

He was right, of course, and I agreed with him. Robert McNamara had once observed how "you must sometimes do evil in order to do good." Everyone in the chain of command leading up to the shot had taken all safeguards possible. What happened, happened. An unfortunate byproduct of waging this kind of war. We had to press on with the mission if we hoped to leave this country in better shape than we found it.

But even knowing that we were right—knowing that nothing we could have done, or could do now, to change what fate seemed to have deemed for those children—did not lighten the darkness that crept through our souls. I telephoned Trish after it was all over. Although that terrible day was disappearing into the approaching purple wine of an Iraqi night, what happened would never vanish from my soul.

"Honey? Matt, is something wrong? You sound . . . funny."

Trish had been concerned that the cough I developed in Iraq on my previous tour would return, but I could tell she knew there was something different. I longed to tell her about it, to spill it all out, to share my revulsion and self-loathing at what we had done because of war. But not like this, not on the telephone. This had to be done in person, at the right time.

"Everything's okay, Trish. I just wanted to hear your voice. Have you been back to the doctor?"

"They're doing more tests," she said after a long pause. I realized she was reluctant to add to my worries. "Matt, they don't think I have rheumatoid arthritis. Matt . . . Matt, I'm frightened."

CHAPTER 21

Out of the Loop

L ONG BEFORE I ARRIVED at Creech Air Force Base for
Predator training, I had pondered the morality of war,
especially the use of "sanitized" warfare in which you seldom
got dirty or had blood splashed in your face. The killing of the
two children, inadvertent though it had been, brought those old
internal struggles back to the surface. Although I couldn't have
kept putting on my uniform every day if I didn't believe in what
we were doing, there would always be doubts.

For one thing, the incident forced me to consider once again
the question every warrior must confront sooner or later: Could I
kill up close and personal were the situation to arrive? Not kill with
a Hellfire missile from the safe confines of a remote GCS where
targets were images on a screen, but instead kill with a bayonet
or a pistol or a rifle. Could I look into an enemy's eyes, recognize
his humanity, and then slit his throat or place a bullet between
eyes that were looking into mine? The only honest answer I could
come up with was, I don't know, but I should think so if it were
necessary and unavoidable.

It seemed that the advance of warfare from clubs and spears
to arrows and bullets and finally to precision-guided munitions
and Predator-mounted missiles had been a progression away
from face-to-face killing of the enemy on the battlefield to killing

from the greatest possible distance as precision and effectiveness increased. With unmanned aircraft, we were on the precipice of redefining the manner in which wars were waged and, at the same time, perhaps changing the texture of human society. The innovation of so-called "smart weapons" was a defining moment in history, ranking with the introduction of the phalanx, the chariot, and firearms. The phalanx catapulted Greece to world power; the Persian Empire rode to glory on the chariot; and Europe conquered the world with gunpowder. The full effects of modern innovations in war were still in flux.

The history of warfare might be considered a history of missing the target while attempting to become more accurate. In 1942 and 1943, it took thousands of B-17 sorties over Germany to make sure of having at least a 90 percent chance of hitting the intended target. Millions of artillery shells and rockets were fired with not much greater accuracy. Thousands of people died in what came to be known as "collateral damage." Today, a single Hellfire missile fired from an unmanned aircraft could do a job that would have required a hundred or more B-17 sorties during World War II.

Strictly from a moral perspective, however, killing people with a bomb or a remote-controlled missile was no different from doing it with your bare hands. Warfare by remote was obviously fresh territory requiring new discussions on the morality of war and spawning the one big question that had yet to be answered: Is it immoral to wage war on humans with automated machines?

"I have worked in artificial intelligence for decades," said Professor Noel Sharkey, University of Sheffield, England, "and the idea of a robot making decisions about human termination is terrifying."

"To be able to carry out operations with less human cost makes sense," declared Peter Singer, author of *Wired for War*. "It is a great thing. You save lives. . . . On the other hand, it may make you more cavalier about the use of force."

People are more likely to support a war if they view an action as costless to their soldiers and if it is conducted in a "surgical" manner. The increasing use of robots offered them a scenario of pain-free military action—at least for those nations with the supporting technology. More and more, armed forces seemed to want to dehumanize military operations and hand over the responsibility of killing to conscienceless machines, thus making death more abstract and less reprehensible. Whether it was a great way to protect our troops or a cold-hearted cop-out remained part of the moral debate.

Either way, fully autonomous decision-making killer robots were high on the U.S. military agenda. The U.S. National Research Council advised that the United States should "aggressively exploit the considerable war fighting benefits offered by autonomous vehicles" that were "cheap to manufacture, require less personnel and perform better in complex missions." The United States was pouring $230 billion into future unmanned combat systems that could strike from the air, sea, or land. Congress set a goal of having one-third of all ground combat vehicles unmanned by 2015.

"At present," said Owen Holland, professor of computer sciences at the University of Essex, "they [robots] require a human to give, by remote, permission to fire, but it will not be long before they can take the human out of the loop."

Although U.S. officials insisted that a human would always be "in the loop" when it came to "pulling the trigger," a Pentagon research project was actually entitled "Taking Man Out of the Loop."

Other nations and even groups within nations were attempting to follow the path of the United States toward deploying remote technology to battlefields, perhaps creating a whole plethora of problems that went even further into the morality issue. Said Professor Sharkey, who also served as chief judge for the long-running TV series *Robot Wars*, "Once you build them, they're easy to copy. . . . We can't really put the genie back into the bottle."

China, Singapore, and Britain were increasingly using military robots. So was Italy, as I had personally witnessed. South Korea and Israel were deploying armed robot border guards.

Iranian-backed Hezbollah guerrillas sent a small remote-controlled drone over Jerusalem. It wouldn't be long, many military people were predicting, before al Qaeda and other terrorists would want to get in on the act and really let the genie loose.

"If you don't really give a toss," Sharkey pointed out, "you can just put an autonomous weapon running into a crowd anywhere. It's only a matter of time before that happens."

On the other hand, removing soldiers from harm's way, replacing them with unmanned war machines, might actually backfire and make a nation that depended on technology more vulnerable rather than less. Somalia was a good example. The United States lost the fight politically after taking casualties and leaving the high ground to a raggedy bunch of "skinnies" who were neither as dependent on technology as Americans nor as adverse to having their own people killed. A news editor from Lebanon spelled it out:

"They [Americans] don't want to fight like real men," he wrote. "[T]hey're afraid to fight, so we just have to kill a few of their soldiers to defeat them."

I was inclined to think that many so-called "experts," with their cynicism expressed against the use of unmanned technology, were not seeing the complete picture. Whether the United States now or in the future becomes overwhelmingly dependent on machines and turns over life-or-death decisions to them is a strategic call not yet made. The Predator, while a modern automated aircraft, was not autonomous. The requirement for human guidance at every step of its operations was its limitation—but also its strength and the reason why the system was so successful. Predator depended on the tactical and technical competence of the crew that flew it.

I believed that Iraq and Afghanistan, with our proliferation there of robots and RPAs, would ultimately prove the value of remote technology. Mere men could not stand up against remote-controlled machines and the determination of the crews that operated them. Fielding a manned presence on the front lines, with all the limitations of a flesh-and-blood fighting force, could never compete with the unblinking eye of a remote presence and the rapid response capability of a kill chain that never quit.

At the same time that that force had become more effective than at any other time in history, it had also become less lethal to civilian populations. Even so, the deaths of two Iraqi kids on the bicycle reminded me that no matter how far we positioned ourselves from the battlefield, robots could never render war completely antiseptic any more than had the crossbow, the tank, or the bomber airplane. Technology changed how men fought and died, might even minimize dying—but it could never change the reality of death.

My resolve for what I was doing never wavered, although I had to admit that there were times when I wondered if I might not be rationalizing for my own conscience. All soldiers must justify themselves if they wished to call themselves honest warriors rather than mercenaries. Collateral damage—the loss of innocent life—was inevitable, no matter what precautions were taken against it. Flying the Predator, firing precision-guided Hellfires that slammed exactly on-target almost every time, contained a moral meaning that might not be apparent at first. It saved our soldiers' lives and, compared to the carpet-bombing of World War II that wiped out entire cities, demonstrated our value of human life and our efforts to do whatever possible to avoid taking it. If we who operated battle machines did our jobs properly, wars would be shortened and fractured societies rebuilt more quickly and securely.

The incident with the kids was unfortunate. Their deaths would plague me. But it wasn't as though we had dropped a string of 1,000-pound bombs to wipe out an entire neighborhood. The two boys, like the old man at the wall, just happened to be in the wrong place at the wrong time. That was how I had to look at it. And that was how I would try to explain it later to Trish.

In the air force, there was an invisible but almost impenetrable wall between intelligence and operations. It sometimes took hours, even days, for actionable information to sift back down through layers of command to a Predator squadron or other field unit. It was a phenomenon we called "stove-piping." Smoke had to travel all the way to the top of the stovepipe before it could escape and find its way back down to the operational level.

Not that the rules had changed; there were simply more people in the command loop, each inserting his own interpretation of the rules. As the Predator enterprise grew, as we flew more CAPs in support of more units, the situation became more complicated. Pilots and crews had had a lot more autonomy when I was flying out of the Nellis POC a year ago, almost as though we *were* flying line of sight.

Captain Bobby Rangler complained bitterly that what we needed was a shorter stovepipe that would give the 46th Expeditionary Reconnaissance Squadron a faster and more direct conduit to work with soldiers on our little corner of the battle-field. Even though my crews took off planes early and landed them late to get in local flight hours before and after the POC at Nellis did its thing, we still faced situations that hampered effec-tively employing those hours to the benefit of troops in the Balad vicinity and the security of the air base.

"Johnny Rico needs to *ride*," Bobby railed.

One afternoon from the LRE at Balad, we watched the feed as a Predator crew at the Nellis POC followed a truck full of rifle-wielding insurgents all over Baghdad. Eight of them. I thought of the suicide bomber who, a week ago, blew up thirty-four small boys scooping up candy tossed from an American Humvee, killing all of them. My one thought about the truck and its occupants was, Shoot the bastards. Every one of them we could find. But by the time we contacted a JTAC and wrung a decision out of the stovepipe, the truck had arrived at its destination and the insur-gents had scattered.

Bobby and Falisha Rexford, his sensor operator, were livid with frustration. Bobby suggested that somebody up the chain must be getting gun-shy to increased public scrutiny of Predator operations and the few isolated incidents in which bystanders were injured. For all we seemed to be utilizing Predator and for all the good we were doing, he grumbled, we might as well turn it into a nationally televised water pistol.

A marine aviator in an F-18 could bomb the hell out of a city block in order to knock out two snipers—and that was war. Depending upon who might be watching our feed, any Predator that clobbered a truckload of insurgents armed with mortars and rockets and unintentionally scratched a washerwoman hanging out laundry nearby became tantamount to a war crime.

A few days after we lost the truckload of insurgents, I was wrapping up paperwork at the end of a shift when Bobby called me to his GCS.

"Matt, maybe you oughter come on down here. We got some ol' boys that need dusting."

Bobby and Senior Airman Rexford had control of an aircraft for the hour after the Nellis POC turned it over to them for landing. On the way back to the airfield, they spotted a five-man mortar team running along a canal bank, apparently spooked by U.S. Army helicopters. By the time I reached Bobby's GCS, the bad guys were wading waist deep across the canal. They were in a hurry.

They scrambled through reeds up the opposite bank to a road, where a Toyota pickup rushed up and slid to a halt. The two men tossed their tubes into the back and piled in on top of them, and the Toyota took off in the direction of Balad.

Shortly, they switched from the pickup to an SUV, which hauled them to an isolated spot on the banks of the Tigres River. An outboard boat was tied up waiting for them. They cranked it up on a plane and roared down the river until they finally reached a safe house overlooking the water.

At almost any point along their flight, we could have nailed them—except I couldn't get permission from the stovepipe to fire. By the time I finally obtained necessary clearance to at least sparkle a fighter plane on-target, the bad guys had cleared out of the safe house and we lost them in the city.

"Fu-uck a du-uck!" Bobby bitched, breaking single-syllable words into doubles to better express his displeasure.

Bobby and Rexford were made for each other as a team. Like him, she was bright and ambitious and wanted to learn everything about the Predator, including how to fly it. Captain Rangler all over again, only female, prettier, and without the baggy flight suit and Okie "tawk." Bobby was obliging. He taught her how to use the radio and taxi the aircraft. Her next step was to actually fly it.

Falisha began to keep a detailed logbook of each day's flights, charting in it dates, times, and locations of insurgent activity as well as any other normal or abnormal occurrences that might provide a broad picture of what was happening within a ten-mile radius (mortar range) of the air base. She was like a cop who knew her district so well that she could tell immediately when something was out of place.

One afternoon while she and Bobby were cruising over the town of Balad, she spotted two strangers in a boat chugging up the Tigris River. She brought it to Bobby's attention. They followed the boat to a weapons cache concealed in thickets along the river. The situation warranted a shot, but again I couldn't get a quick decision for them. The army brigade in charge of the area around the base wouldn't clear it until they saw the video and confirmed that the target was indeed hostile. That meant piping the video feed to the army's tactical operations center (TOC) so the battle captain could view it and decide on a course of action. By the time all that occurred, the bad guys had stashed their load of weapons in the thicket and were clearing out of the vicinity.

The bad guys got away because of stove-piping. I couldn't even convince them to send a team out there to retrieve the weapons. Bobby and Falisha were more frustrated than ever.

One of the proposals in the integrated base defense concept of operations that I was working on, along with the line-of-sight suggestion, was that we streamline the kill chain so that the Predator crew identified the target and determined if it were hostile or not. That way the army battle captain could clear the shot without necessarily having to see the live video—just like a forward artillery spotter. Bobby's and Falisha's weapons cache and the insurgents who escaped incident went into the report as yet another example of why the stovepipe needed to be shortened and what we might accomplish if it were.

Everything depended on Gen. Frank Gorenc and his recommendation to the CAOC once I presented the completed report to him. He was still eager to take over Predator controls and fly a mission himself. He showed up to fly; I demonstrated a full handback with mission coverage, followed by an approach and landing. I had to cut the flight short when a dust storm blew in.

"Tomorrow then, Matt?" He threw a massive arm over my shoulders. "So, how's the little woman coming along?"

I hadn't told him of Trish's medical problem. I rarely spoke of my private affairs, preferring to compartmentalize my life as not to end up like Brent obsessing over home and unable to properly work our mission in Iraq. The general, however, had somehow learned of her condition. He made it his personal business to know everything about the people under his command, one trait of a good leader.

"She's undergoing more tests, sir. We'll know something soon."

"Fine, Matt, fine. If there's any way we can help . . ."

I turned over Predator's controls to him the next day while I stood by to supervise. It was like training a new first lieutenant to fly, except my student in this instance wore stars and was a general officer. It was exhausting for me, exhilarating for him—and worth it all.

"Sir, there is *something* . . ." I ventured after he landed.

I gave him a completed copy of my CONOPS and a short briefing to go with it. I showed him the tapes of Captain Brent's capture of the gunmen in the white pickup and of the insurgents in the boat that led Bobby and Falisha to the weapons cache, and then escaped due to stove-piping. I also showed him Airman Rexford's notebook and how my concept, if accepted, would provide increased tactical coverage and protection for the air base. We had discussed all this before, but not in such depth and not with a report to back it up.

"If we had rules of engagement in place that allowed us to coordinate immediately with our people on the ground in a local line-of-sight operation, these insurgents wouldn't be getting away from us to continue their attacks against our forces," I explained. "This is a way to take advantage of time on my schedule that's not being used. It would be easy to set things up for immediate tactical action."

He nodded. "I told you before, it's a good idea," he boomed. He tapped the report with a thick forefinger for emphasis. "Now we got something to go with it. Get the army on board, Matt. I'll send up your report with a recommendation. Let me know if you need anything else."

CHAPTER 22

Neighborhood Watch

MAJOR WILLIAM "SLACK" ROBERTS, the officer I relieved as commander of the 46th Expeditionary Reconnaissance, and with whom I had previously served at Nellis, thought the best way to motivate people was to smack them upside the head. Not literally, of course, but it was the same thing. He was like Mr. Moran back at the Predator formal training unit and his constant diatribe against student "shitheads" and "numbnuts." In keeping with his smacking theme, he had initiated a "Dunce Award" presented to those on his staff who committed boneheaded acts of stupidity or carelessness.

I kept the tradition for a few weeks, Captain Brent a frequent "Dunce" winner, until I realized that negative reinforcement was not my leadership style. I replaced the Dunce with a "Hoo-ah Award," named for an energy bar served at the chow hall and for the enthusiastic response that warriors made to almost any question or comment: "Hoo-ah, sir!"

I placed a fish bowl on my desk into which crew members could drop the names of those whom they thought had contributed positively to the mission through achievement and good attitude. The winner each month received a Hoo-ah bar and his name inscribed on a makeshift plaque on the HAS wall. The Okie replaced Brent as the most frequent winner of the award, followed closely by Captain Henson—until Statler rotated back to the States.

Captain Clark Hurst replaced Statler. He was about thirty or so, average in height, serious, professional, and rather quiet. A fine addition to the team and a monthly contender for the Hoo-ah.

One morning shortly after my impromptu conference with General Gorenc, I walked in and dumped a double handful of energy bars in front of Bobby and his sensor, Falisha Rexford.

"As far as I'm concerned," I exulted, "you two have won the Hoo-ah of all Hoo-ahs. Thanks partly to your notebook, Falisha, and the tape the two of you burned on the weapons cache by the river, we're being given a chance to prove ourselves for a line-of-sight base defense mission."

"Shucks," Bobby drawled modestly. "It was the brilliance of your CONOPS that did the trick."

Although some in the CAOC were apparently trying to impede the reinstitution of a local security mission flown by my pilots, General Gorenc was determined to help me push it through. I got the army on board, as he suggested, by taking a copy of Bobby's and Falisha's tape to the 1st Brigade Combat Team of the "Big Red One" 1st Infantry Division, the unit responsible for the AO that included Balad Air Base and the surrounding area. Major Doug Winton was the brigade S-3 director of operations and, coincidentally, the battle captain at the TOC the afternoon that "stove-piping" allowed Bobby's bad guys to cache their weapons on the riverbank and escape.

Winton sat down with me in front of a computer. I saw his face as he watched the two insurgents unload an 82mm mortar tube and ammo from their boat, carry it up a path through the reeds to where they cached it, and then stand around like innocent children while army Apache helicopters flew overhead. His jaw dropped.

"You must have been pissed at me," he observed. "I was the one who refused to let you kill these guys. But by the time I saw the video, the show was over and the Hajjis were gone."

I had indeed been positively livid at the time. But that was in the past. Working together now with the army, we quickly formulated the framework for a new way of conducting base defense, beginning with a clearance-to-fire drill that allowed army Major Winton and his fellow battle captains to rapidly assess a situation and make the call even though they might not be looking at the video. Integrated into other base assets—Apache helicopters, F-16s, army cameras, and radar fixed around the base—Predator would occupy the center of the operation.

In order to ensure proficiency, we devised coordination exercises and drills until we could go from "eyes on" of an enemy to "cleared hot" in under one minute. It seemed we were finally going to do something to stop the daily attacks and convince insurgents that it wasn't worth the risk to continue launching mortar rounds at the base. Everything now depended upon how well we conducted ourselves in a Predator base defense exercise, a mission rehearsal approved by General Gorenc to test my proposals. Thanks to the general and to my hard-working crews of pilots, sensor operators, maintainers, and support personnel, I didn't see how the CAOC could turn us down.

The concept of the exercise was to begin with the Predator crew receiving intelligence about an enemy mortar team played, unsuspectingly, by IA (Iraqi Army) soldiers in a truck. All they knew was that they were ordered to show up at a particular Baghdad intersection at a certain time, remain static there for another period of time, then travel across the city to the outskirts. One of my airborne crews would coordinate with tactical air control and a quick reaction force to either "kill" or "capture" the bad guys at some stage of the scenario. The objective was to demonstrate how much more rapidly a well-coordinated, well-integrated Predator-centric operation could control and resolve a situation if the "stove-piping" chain of command was eliminated by having us work directly with tactical elements. I had a lot riding on the outcome, possibly even my job.

Just to make things more interesting, everything was to play out in the middle of the night. Excitement reigned over the Johnny Rico Squadron.

As commander and the officer most responsible for the mission, and with the most to lose, I chose to fly the airplane with my sensor, Senior Airman Kimberly Joan. I selected *Ruby* for the job and made special arrangements with the Nellis POC to have the airplane's flight extended to cover the rehearsal. Everything was a go.

"Catch some sleep, sir," Kimberly clucked. "We need you fresh and alert tonight. You understand?"

"Yes, Mother."

I telephoned to check on Trish before forcing myself to nap before the test kicked off. She was in pain but bravely assured me that she was doing okay.

Brent, Bobby, the new pilot Clark Hurst, and their sensors crowded into my GCS to watch as nightfall spread across the runways at Balad. They were in clear violation of crew rest rules, but no one would have been able to sleep tonight anyhow, not with so much riding on the outcome. A couple of army imagery analysts mixed in with my crews and stood behind me in the GCS to render assistance. General Gorenc would be watching from his own video feed.

Right on time, *Ruby* returned from her Nellis mission with plenty of fuel to cover the exercise. I took control, checked in with the JTAC by radio to receive the target's grid coordinates, and arrived over the correct intersection less than fifteen minutes later. The entire mission would be flown through the infrared camera, which picked up radiation heat signatures and transferred them into images on my screen.

"Let's go get some!" Bobby cheered.

A curfew was in effect, which limited traffic on the streets to military, Iraqi police, or other emergency vehicles. Insurgents, of course, were still scurrying around like rats in the shadows of alleys and gutters trying to pick off a stray patrol.

The IA truck was not where intel said it should be. All I saw was a U.S. Army patrol passing through in Humvees. Bobby groaned. Hurst muttered something. Brent was already sulking. Screwed before the test even began. Still, as we say in the air force, flexibility is the key to airpower.

Exasperated but nonetheless resolved, I scouted adjacent intersections. An F-16 fast mover participating in the coordinated exercise also reported his inability to locate the target. You could have cut tension in the GCS with a butter knife. If the 46th failed, we would all be landing and taking off airplanes for the duration of our tour, feeling like spectators watching the war go by.

I picked up a heat signature shrouded in the shadows of some palms three blocks away. Closer inspection revealed it to be our truck. I couldn't help wondering if the mix-up might not have been deliberate sabotage by certain higher-up officers who, for reasons of their own, wanted the mission to fail.

Kimberly locked on to the truck with her sensors. No way were we going to lose it now. Not tonight.

We tracked the target from the air as it started up and began to twist and dodge through the darkened city en route to its next stop. Our mission was actually quite simple once we located the truck: Guide an army QRF to the target vehicle by radioing directions based on the video of my one screen and the electronic map of the Balad area on the other.

"Don't muck it up now," Bobby encouraged.

The IA truck stopped to wait at its next assigned location on the outskirts. By now, the QRF was less than a half-mile away, following my directions. On my cue, Kimberly fired up her target marker and shot an infrared laser beam onto the street about ten meters in front of the QRF's lead vehicle. Night vision goggles made the beam clearly visible to the Humvee driver. All he had to do was follow the beam. Sort of like leading a donkey with a carrot. Down one street, up another—until I sparkled the target and the QRF surrounded it. It was all over within minutes.

It was a huge success, aside from the initial confusion. Almost too easy. A crew at the POC in Nevada could have done the same thing. In fact, *I* had done it before. Where we trumped Nellis was in being locally integrated with all the various systems and assets. Pilots or mission commanders at Nellis lacked the capability to drive across base and personally interact with army teams, QRFs, JTACs, and other commanders. Stove-piping slowed them down, sometimes prohibitively. We on-site guys were the answer when the defense required speed and prompt attention.

The night's exercise validated the tactics proposed in my CONOPS. If the CAOC accepted it, we wouldn't lose as many bad guys from now on.

General Gorenc burst into the GCS. He had watched the screen in Ops throughout the exercise. He wrapped a big arm around my shoulders.

"You proved it, Matt!" he roared. He swept his other arm to indicate the rest of the team. Everybody was grinning, even Brent. "All of you did. They'll have to sign off on it now."

Operation Neighborhood Watch would soon be under way.

Bobby hitched up his gun belt. "Johnny Rico rides again."

Predator was a slow airplane. Approach speed to landing was between sixty-five and seventy-five knots, a sparrow's pace compared to that of the F-16s, with which we shared the pattern. Since Balad was the busiest airfield in Iraq, we had to take up as little time as possible on the runway when we launched or recovered an airplane. I developed a habit of landing on "brick one"—on the numbers at the beginning of the strip. I then cleared the runway by exiting on the first high-speed taxiway in order to prevent the aircraft stacked up behind me on final approach from having to pull throttle or go around.

During landing, the Predator's nose camera remained fixed in the forward position so the pilot had a steady visual reference of the runway and the plane's attitude. Air controllers knew to keep the first taxiway open since RPAs could not see down it prior to turning off the runway. One afternoon, things didn't work that way.

I landed *Ruby* on "brick one" as usual and rolled onto Golf taxiway toward the hangar. I knew something was terribly wrong when "copilot" Kimberly Joan emitted a sudden blood-curdling scream. "*Stop!*" Her targeting pod camera provided a slightly wider field of view over my nose camera and therefore a split second's advantage in seeing a threat.

The next instant, a C-12 transport filled my screen, beak to beak with my plane, its twin turboprops turning, two pilots and six passengers aboard. In a moment that seemed to stretch into forever, I clearly envisioned the inevitable—*Ruby* careening into the C-12, the C-12's propellers chewing through Predator's Kevlar skin, engine disintegrating; the C-12 collapsing inward as the nose of the Predator crashed its way into the cabin, pilots and then passengers being crushed, and both craft exploding in a fiery field of debris.

Training and thousands of flight hour experience took over. Instinctively, I slammed on the brakes by pressing on the top of the rudder pedals and watched the nose of the aircraft drop as she started to slow down. Not good enough. A collision was still imminent. Again by instinct, I stomped the left rudder pedal to swerve my airplane to the left. I came to rest perpendicular to the C-12, my right wingtip within a few feet of its spinning number two engine. I drew in a long, wavering breath of relief, almost collapsing at the controls from the horror of the near miss. Tower had failed to keep the first taxiway open and then had further exacerbated the situation by not informing me of it. A Predator could not see around corners.

Still, I was not totally without fault. Familiarity and repetition had led to complacency. Plus my mind had not been completely on the moment; home and Trish's health preoccupied it at times when I needed to be fully alert.

During most previous wars, communications between the front and home took weeks via the army post office (APO) and the U.S. mail. Tattered letters with news of family crises, births, marriages, deaths, and neighborhood gossip reached soldiers, marines, and airmen in rather irregular "mail call" fashion. By the time a serviceman received his mail, whatever had occurred was already in the distant past and out of his hands. There was nothing he could do about it.

War had now gone modern with Internet and the satellite telephone. Although ready communications provided family contact on a regular basis, not much different from communications between family and a soldier who was on a trip to a neighboring city, it could also be a double-edged sword. Not only did the serviceman have to deal with the war, he had to cope with family matters often intensified by his absence. A divorce filing, a sudden illness, a kid going to the dentist, in-law conflicts, the death of a pet, rebellious children. No wonder Captain Brent was sometimes a human wreck.

I telephoned Trish at least twice a week and e-mailed her almost every day. She sounded withdrawn, at times upset. I knew she was holding something back and I knew why. During my deployments, she always strived to handle the home front on her own rather than burden me.

Finally, upon my insistence that she tell me everything, she informed me that doctors had completed their tests. Her condition had been misdiagnosed.

"It's not rheumatoid arthritis. It's . . ."

There were tears in her voice. I immediately assumed the worse. "Trish!"

"Matt, I have lupus," she blurted out. "We're starting chemotherapy tomorrow."

I didn't know what lupus was. "*Cancer?*"

"No, no, Matt. It's not cancer. They think they caught it in time to control it."

I didn't like the sound of that. "*Control* it?"

"There's no cure for it. I'll have to learn to live with it. Oh, Matt . . ."

"Honey . . ."

I had never felt so much distance between us; I was halfway around the globe from her.

"Matt, I wish you could be here with me."

I looked up systemic lupus erythematosus on the Web and learned that it was one of several diseases known as a "great imitator" because it mimicked or could be mistaken for other illnesses, such as rheumatoid arthritis. It was a chronic auto-immune connective tissue condition in which the immune system attacked the body's cells and tissues, causing inflammation and tissue disease. It occurred nine times more frequently in women than in men, especially in women between the ages of fifteen and fifty. Symptoms included fever, malaise, joint pains, myalgia, and fatigue, all of which Trish had experienced over the past year.

Although incurable, sometimes even fatal, it could be held in abeyance with corticosteroids and immunosuppressants. Multiple genes appeared to influence a person's chance of developing lupus. Environmental factors such as extreme stress could trigger an episode. I felt arrows piercing my heart. Were I and the war at least partially responsible for my wife's condition?

"I can't handle things by myself anymore," Trish finally e-mailed. "The chemo and the medications are making me crazy. Matt, you have to come home. I need you."

I telephoned. The tone in her voice . . . she was crying. I still had work to do in Iraq, but Trish was my life.

"I'll talk it over with Colonel Jacobs," I promised. "It's going to be a bit of a trick, honey."

Colonel Jacobs, the new Ops group commander, empathized, but he sounded pessimistic about ending my tour early. After all, I was a squadron commander; mission came first. I knew how things worked. I sat on the board that reviewed cases of airmen requesting to be sent home for personal reasons. We rejected more appeals than we granted.

I was starting to sound more and more like Captain Brent. Weeping on Colonel Jacobs' shoulder. Almost running into a C-12. I had to pull things back together, for the squadron and my people as well as my own good.

I spoke with Trish's doctor, who reassured both of us that she was not likely to die from this condition. Treatment in its initial phases might cause a certain amount of disorientation and emotional stress, but she would settle down and be all right once the medication took hold and leveled out her metabolism. Until that point, it was best that she not live alone.

I contacted my cousins, Ed and Cheryl, who also lived in Las Vegas. They offered Trish the spare room in their house. She had been feeling so alone. A little help from loving relatives was what she needed to keep her sane and on track with her treatment.

It was morning in Nevada the next time I called.

"Matt, I'm so sorry to have worried you when you had so many other things on your mind."

That was my old strong, brave Trish, after whom I had named a warplane.

"I know you can't get home until rotation, Matt, but that's okay. The doctor says I'll be fine. Do what you have to do. I'll be waiting for you when you come home."

It took only a few days to weather a crisis that would have taken weeks, maybe even months, for my father's generation at war to have handled via mail call. Technology could be an unbelievable social asset outside my world of RPAs, Hellfire missiles, and kids slain while riding their bicycle.

CHAPTER 23

This Changes Everything

FOR THE PAST TWO months while I was busy soap-boxing for a local base security mission, insurgents kept pounding Balad with mortars. Sit down for chow and *Alarm Red* sounded. Walk to the PX and the siren went off, followed by the thumping *Whump! Whump!* of 82mm or 120mm rounds landing somewhere in Mortaritaville. I would duck into the nearest bunker and seethe with rage while I waited for *All Clear*. Bobby, Clark, Brent, and I plotted for hours on the best way to take out the mortar men who seemed to taunt our every effort to retaliate.

"We are going to kill them," I vowed, pacing. I wanted revenge. I craved it. This had become personal.

What with only a flight hour or two a day purloined from the Nellis POC, we didn't have enough time in the air to dedicate ourselves to catching our tormenters. Silence from the CAOC on the CONOPS I submitted to change all that was like a mockery.

"The army can't catch them either," Brent rationalized in his best "Oh, well" manner.

Finally, it came through. General Gorenc was true to his word. My CONOPS for Operation Neighborhood Watch, the Predator-centric plan for the defense of Balad Air Base, was approved by the Combined Forces Air Components commander, Lieutenant General Buchanan. Everything I had worked for was now on the line. Important people were watching, waiting for us to prove the point of my CONOPS—or screw it up. *If Predator can't even get rid of a couple of scroungy mortar men sneaking up to hurl shells, then what have you been braying about?*

I received approval on August 31, 2005. We commenced operations the next day. At last we had our own aircraft, its call sign Watchdog Four-Six.

Again, Senior Airman Rexford's notebook came in handy. Bobby got up and strode to a wall map of the base and surrounding country, hitching up his flight suit as he went. He and Rexford had marked *X*'s on the map to designate locations from which we had received fire or they had noticed enemy activity.

"They never strike from the same exact place twice," Bobby pointed out, "but they're shooting from the same general area. It's like they're studying our ground patrols, timing them, and then striking from wherever they're not. Bang! Bang! Then they're gone before air can vector in on them, or the base sends out a QRF. That's how we're going to get them."

I traced on the map with my forefinger how the mortar men kept switching from one place to another in response to increased security in an area. That gave me an idea.

The next day I met with Doug Winton, the army director of operations, and explained what I had in mind. Could he position his patrols to demonstrate presence in one area and absence in another?

"Roger that."

The insurgents should, if all went well, notice the difference and attack us from what they considered a safe zone. We would be watching from the sky like a bird of prey ready to strike.

"Sounds like a job for Johnny Rico," Bobby approved.

The dull thud of exploding mortars followed by the warble of *Alarm Red* sirens jarred me from my bunk. Balad was under attack. I glanced at my watch: 0400, the bewitching hour. Watchdog Four-Six was airborne with Bobby and Rexford in the cockpit. They would be hunting.

I hesitated to call the POC, not wanting to distract the operation. Finally, I could contain myself no longer. I rang the number. Senior Airman Mata in Ops answered.

"Nothing yet, sir. I'll let you know."

I returned to bed but lay on my back awake, staring into the darkness, listening. Four days had passed since we initiated Operation Neighborhood Watch. The insurgents were proving to be as slippery as river eels. Despite our best prediction about how the enemy would behave, we either didn't have a plane in the air when we were mortared or the bad guys were gone by the time we got there. Tonight seemed like another one of those occasions.

After a while, when nothing happened, I drifted back to sleep. A short, sharp blast rattled the trailer walls. My eyes popped open. I had heard a lot of explosions in Iraq over the past months—mortars, rockets, artillery, bombs, controlled detonations—but none of them carried a signature like this one. It had to be a Hellfire.

The phone jangled. It was Airman Mata. "Sir, Papa missile ops check good."

"We got 'em?"

"Yes, sir. We got 'em good."

Now was the time to be excited. It was 0445 hours. I hurriedly pulled on a fresh uniform, although I had to remain inside Bobbys as *Alarm Red* had gone back into effect. I paced the tiny trailer, waiting for the *All Clear* so I could rush to the GCS and get the skinny from Bobby and view the tape.

Alarm Yellow sounded after about ten minutes. Personnel with essential missions were allowed to venture out. My trip to the flight line couldn't be considered essential. I fretted out the delay.

A second explosion jarred the trailer. Another Hellfire. I dialed Mata.

"Are we still shooting?"

"We just went Winchester, sir." That meant Watchdog had expended both missiles.

Screw it. Red, yellow, or green, I was out of here and on my way. General Gorenc was already on the flight line, grinning that big Slovenian grin of his and bear-hugging everyone.

"Matt, we've proved your CONOPS. Not once—but twice. Let them doubt it now."

Everything had gone perfectly according to plan. Army's roving security had deliberately left Areas 200C and 200D open on the north side of the base for Predator to watch. Army radar immediately picked up the POO (point of origin) of the first attack in 200C. Bobby was patrolling above only seconds away. Through infrared, he and Rexford spotted two men hurrying away across the field, one of them carrying a mortar tube on his back.

While the Okie sweated out the analyst's confirmation of the target, the two terrorists linked up with four others in front of a house at the edge of the field. One of them had brought a shovel in order to bury the evidence. They tossed the mortar tube underneath a tree and, apparently feeling safe, stood around smoking and joking and undoubtedly recounting their heroic exploits.

Strict ROEs mandated that a Predator could not fire unless there was positive identification of a weapon, accompanied by hostile intent or action, and with the additional stipulation that collateral damage be restricted. The two analysts standing behind Bobby in the GCS decided that the target qualified on all counts. There was no delay in Bobby's receiving a clearance to fire.

"Cleared hot!" the JTAC radioed.

General Gorenc and I watched Bobby's video. There was a slight flash on the screen as he fired. Twenty-five seconds later, the Hellfire streaked into view on the monitor and plunged straight down at the tree under which the five mortar men were contemplating burying the tube. A bright flash washed out the video.

A few seconds later, the screen cleared. We watched one guy stumble away from the splintered tree and collapse on a dirt road. The other four were, as Bobby put it, "assuming room temperature."

"That's worth another Hoo-ah," I praised my pilot and his sensor operator. "You keep getting Hoo-ahs, Bobby, maybe you'll fill out your flight suit."

General Gorenc slapped me heartily on the back.

It wasn't over yet. There was still one Hellfire left.

Minutes after Bobby and Falisha cut the first group to shreds, a second insurgency team loosed an 82mm round from Area 200D. Doug and his people working with their intel and with us had so accurately predicted the next enemy shot that one of their security cameras was pointed in the direction of the attack. The army battle captain at the TOC actually observed insurgents fire the mortar.

Bobby had eyes on the POO within seconds. Five men were scurrying away with the tube. Analysts in the GCS immediately confirmed the target. The battle captain completed his clearance to fire before Bobby could line up a shot.

"Cleared hot!"

They were a band of amateurs down there, the insurgents, all bunched together like chickens as they scurried away. Zarqawi or one of the other leaders was probably paying locals a couple of hundred bucks or so to go out and pop a round now and then at the air base. Until tonight, they had probably looked upon it as a lark, a great way to earn cash that carried no consequences.

None of the five would be shooting again. As soon as they paused for breath along a tree line, Bobby let them have it. The Hellfire left a smoking crater surrounded by body parts. One guy was still alive, at least for the moment. I almost felt sorry for the poor bastard. All broken and bloody, he reached out one hand and clawed at the ground, trying to pull himself along. The other arm lay limp and twisted against his side.

Nobody in the GCS uttered a word. We merely watched the video, mesmerized, both awed and horrified by the carnage and this guy's dying effort.

People from a nearby house ran outside to investigate. Two men grabbed a torso and dragged it to the house but left it outside. A pickup drove up. The driver and the men from the house tossed corpses, body parts, and the one possible survivor into the back of the truck in a grisly pile. The truck sped off with its bloody cargo.

Bobby's recorded video ended and he switched back to real time. Crowded around his monitor, we watched as he followed the pickup about twelve miles to a hospital in the town of Baqubah. Hospital attendants ran out with gurneys. They unloaded the one guy, who seemed to be still alive, and rushed him inside. Somebody tossed a tarp over the mess that remained.

A few minutes later, a car showed up from the first strike carrying one wounded and four dead, all of whom were taken inside. Bobby circled Watchdog overhead until the U.S. Army arrived to arrest the two wounded men and count and identify the corpses.

It was a clean sweep. Within a matter of a few minutes, Bobby had wiped out nine enemy fighters and wounded two more, a record. None of us had ever destroyed so many in a single effort.

My concept had worked better than I thought possible. General Gorenc proclaimed me his hero.

"This is exactly the result we needed, Matt," he boomed. "This changes everything."

Somehow, watching it all on the monitor, my lust for enemy blood dissipated—but not my resolve to see the war through.

The next day, following our big victory over the enemy mortar crews, I faced a challenge of a different sort. I called it the Great Kimchi Caper.

Captain Roger Brent straggled into Ops with a Styrofoam container of fermented kimchi from the chow hall. The stench permeated every corner when he popped the lid. Clark Hurst and Bobby got up and moved to the other side of the room. They had complained of his addiction before.

"That's some rotten shit, man," Clark protested. "Kee-rist! How do you eat that stuff? It's enough to gag a maggot."

The squadron maintenance officer, Lt. Jeff Kuhn, happened to walk in. He froze in the doorway as though slapped in the face by the stink. He was not nearly as tactful as Clark and Bobby in his assessment of Brent's culinary inclinations or in the origin of the material in Brent's Styrofoam cup. Brent took offense; he was seldom in good humor anyhow. The argument heated up and spilled over into my office down the short hallway. Clearly, stress and close Bobbys were beginning to wear on people.

Seldom had I needed to raise my voice in an official capacity. It wasn't my style. But something had to be done to restore and maintain discipline. Breakdown of a unit often began with petty incidents not properly handled.

"You two! Get in here!" I shouted in my best drill sergeant voice.

They knew I meant business. Both scrambled into my office and instinctively snapped to attention in front of my desk. I paused for a minute to let the tension grow, then slowly leaned back in my chair. My voice lowered to a measured tone.

"What seems to be the issue, gentlemen?"

Naturally, I knew. Brent still clutched his cup of kimchi. It was all I could do not to vent my own assessment of his dietary habits.

"Major, ventilation in here is terrible," Lieutenant Kuhn said, speaking first. "His daily kimchi routine is about to drive me and my maintainers nuts. You can smell that stuff all the way out to the hangar."

I paused again and shifted my gaze to Brent. "Roger, what do you have to say?"

"Sir, I don't intend to upset anyone. I just like kimchi. It helps me relax."

I rolled my eyes toward the ceiling to drive home the pettiness of the whole argument. For Pete's sake, these were air force officers. Still, a good leader should make both parties of a dispute walk away feeling that it was resolved fairly, no matter how trivial the disagreement. I settled on a compromise to make each accept responsibility for the breakdown while at the same time making each feel that he was coming out ahead.

"Captain Brent, from now on you will eat your kimchi in the

chow hall. That means you'll have to leave for lunch a few minutes early so you can get back to start your shift on time. Can you do that?"

"Yes, sir."

"Lieutenant Kuhn, if you have a problem with one of my crews, you need to come to me before tempers get out of hand. You should know better than to let your troops see you lose your cool. Can you do that?"

"Yes, sir."

I could see they were both embarrassed by their behavior.

"It's settled then. Get out of my office. Can't you see I have work to do?"

They turned on their heels and left. And thus the Great Kimchi Caper was resolved. I fanned the air to clear it of the reek of fermented cabbage.

CHAPTER 24

Pogo

MUCH OF FIGHTING A war was dull and routine. Even the dogface soldier out in the blood and the gore and the mud spent less than 1 percent of his time in combat activities. The rest of his awake time went into getting ready to fight, eating, going to the can, bitching, and waiting. I met marines and soldiers who had been in-country for months without seeing a shot fired in anger or having an IED blow a Hummer out from underneath them. The old adage about "hours and hours of utter boredom followed by sudden stark terror" had become almost a cliché among combat soldiers and big-city street cops.

Predator Fobbits were for the most part insulated from the baser horrors of war, even though we might view them on remote video. For my 46th Expeditionary Reconnaissance Squadron, "sudden stark terror" rarely meant the threat of imminent death for a downed pilot, or an IED exploding on patrol, or a sniper's bullet. A mortar round might improbably land on or near one of us, as with the sergeant from maintenance who was leaving the barbershop. On the other hand, a falling meteorite could pop me upside the head, a deuce-and-a-half truck could back over me, or I could stumble and fall going down stairs and break my neck. Terror for the pilots and crews of the Johnny Rico Squadron generally entailed a sudden in-flight crisis and the threat of crashing an airplane.

The Predator is a delicate machine. It has a fragile look to its thin gray wings, slim fuselage, and spindly landing gear. In flight, landing or takeoff emergencies were almost routine. The inescapable reality was that we were going to damage or lose planes from time to time. Since July, Brent had scraped up a plane's tailboards, I had almost collided with a C-12, and Statler crashed a Predator after he had already landed it. The best I could hope for as commander of the squadron was that we minimize our losses through superior risk management and training and, when we *did* damage a Predator, sustain combat capability long enough for a replacement aircraft to arrive.

One of Bobby's wing control modules failed during climb-out from the airport. The plane rolled into a hard left eighty-five-degree bank and stalled before nosing into a steep dive. Earth rushed up at him through his monitor. He lost several thousand feet of air. It appeared the Predator would crash and burn.

With only a few hundred feet to spare, his pilot instincts kicked in and he applied a classic unusual attitude recovery: opposite rudder, increased power, and roll the wings level. Having regained some control, he was able to land the aircraft unscathed. The baggy-pants Okie would have made one hell of a fighter pilot. I recommended him for an Air Force Safety Award.

Somewhat later, the left aileron on my plane stuck seventeen degrees in the up position. Under normal operating conditions, the aileron banked the plane in and out of a turn. Stuck and immobile, however, it acted like a big speed brake that produced more drag than engine power could overcome, thus hampering the aircraft's ability to either turn or maintain altitude. I thought I might have to deliberately crash it.

I contacted the tower, declared an emergency, and requested the nearest runway. Using nothing but rudder and throttle, I flew a wide pattern to line up on the runway. Fortunately, there was no crosswind. In fact, there was almost no wind.

I knew I had to come in steep and fast to make it work. I would get only one shot at it since I had to use full power simply to slow down the loss of altitude.

Even under normal conditions, landing an RPA was more difficult than landing a conventional aircraft. If you were actually inside a cockpit, you saw the attitude of your plane through direct and peripheral vision, heard the spool-up of the engine, felt the ground rush, and knew instinctively when to flare and ease the airplane onto the runway.

None of that applied to Predator. You looked through the video camera "soda straw" with almost no peripheral view. You saw only what was square ahead, your view fixed to the nose. If you had to crab into a crosswind, your view crabbed with you. You couldn't hear the engine, couldn't feel the ground rush. Everything was pure instrument and video interpretation. Add to all that a microsecond control delay due to the electronics relay between GCS and plane and your timing could be thrown off.

Bobby and Falisha, some of the people from Ops, and even General Gorenc crammed into my GCS to watch me work and, perhaps, make bets on the outcome, with the odds not in my favor. Kimberly Joan sat frozen at her station next to me.

I lined up on final in anticipation of a steep descent. I pulled back power just a bit, knowing it wouldn't take much reduction to drop the aircraft like a rock. I focused on pitch to maintain airspeed and glide path, waiting for the right moment to flare.

The plane sank toward the runway and into the ground effect caused by the cushioning of air vortices bouncing off the asphalt. I applied back pressure to the control stick and added a touch of power to arrest my descent and set up for the touchdown. So far, a nice, smooth glide.

At the right moment, I chopped power and factored in a touch more back pressure on the control stick, bringing up the nose slightly. As soon as the main landing gear kissed the runway, I gently lowered the nose gear to a smooth and uneventful roll-out. The little crowd around me cheered. I could almost hear Moran back at Nellis saying, "Not bad, Shitbird. You finally got your head out of your ass."

"Good work, Matt," General Gorenc said.

He let the other shoe fall after I turned the Predator over to the ground crew and stood up to unwind.

"Matt, they're thinking about grounding the fleet if the malfunctions continue."

"Sir, you can't let them do that. Would they ground the F-16s?"

"Good point, Matt. I'll bring that up. In the meantime, you and your people are doing a great job."

That was the last I heard about redlining our planes.

In the meantime, the war continued around us at its own pace and tempo in spite of our routines and little dramas, among which were malfunctions, Brent's kimchi and homesickness, and Bobby's baggy flight suit. Nearly a thousand people trampled one another to death in a mob stampede on the Al-Aaimmah bridge after rumors of a suicide bomber caused panic among pilgrims to the shrine of the Imam Musa al-Kazim. And in Baghdad, Iraq hanged three men in the first executions in the country since Operation Iraqi Freedom began. They were convicted of kidnapping and beheading three policemen and with abducting, raping, and murdering Iraqi women.

It was rumored that Saddam Hussein might be next to meet the hangman. If I had had a say, I would have selected Abu Musab al-Zarqawi as the next nominee for Noose of the Month Club—as soon as we caught him. Followed by Osama bin Laden—as soon as we caught *him*.

As for the general riffraff and rubble in the world of terrorism and insurgency, they still seemed not to have a clue as to how we knew so much about them and their activities. Mortar attacks on the air base declined significantly after Bobby and Falisha wiped out so many gunners in one night. Operation Neighborhood Watch was putting a major crimp in the enemy's mortar business—and they apparently couldn't figure out why.

After hours and hours of airborne surveillance, not once had I observed a single insurgent with a tube or visiting a weapons cache who so much as looked up to scan the sky. It was almost like they were ants down there, predictable in their behavior to some degree of mathematical probability, no more aware of Predator's presence than they were of the Almighty watching them. In fact, they may have been more cognizant of Allah than of Predator.

Allahu Akbar!

For two hours one night, Bobby and Falisha watched two men digging holes in a field. They appeared to follow no particular pattern; they certainly weren't cutting a drainage ditch or an irrigation canal. They would dig for a while, examine the hole, refill it, then move over a few yards and start all over again. Bobby and his sensor operator thought they might be looking for a dislocated weapons stash.

They continued to observe the men's odd behavior until they ran low on fuel. Bobby directed a QRF to the site before he turned toward home. Cornered, the Iraqis claimed to be digging for water. In the middle of the night? They were not detained.

It was a warm, bright morning with miles of visibility. As Watchdog Four-Six on a random surveillance of the base, I picked up a fellow walking across a field with a determined stride. Since I had nothing more pressing pending at the moment, I decided to keep tabs on him for a while, see what he was up to. Any kind of suspicious or unusual activity rated at least a few minutes' look-see.

He walked up to where a small grayish mule was browsing in the field. He stopped to look around before he looped a rope around the animal's neck and tied it to a shrub. The guy lifted his man-dress, approached the mule from the rear, and, without further foreplay, began to service it. I called Bobby over to take a look at the proceedings.

"What, he thinks it's a sheep?" Bobby cracked.

"Isn't that how country boys get their loving in Oklahoma?" I cracked back.

"Only if the sheep is pretty. And you're in love with it."

My guy down there had no idea he was starring in his own video, the Iraqi version of *Debbie Does Dallas* with a more perverted twist. Most disturbing thing I had yet witnessed in Iraq.

General Gorenc wrapped a big arm across my shoulders. He wanted me to brief a congressional junket of politicians about Predator's capabilities.

"You need to do this, Matt. Politicians being politicians, they won't understand what you're saying anyhow. It's an opportunity for them to go back home to their constituents and boast how they've been to Iraq."

He chuckled. " 'We have met the enemy and he is us,' " he quoted.

"Sir?"

"Pogo."

"Pogo, sir?"

"It's an old Sunday Funnies comic strip."

"Like politics, sir?"

We both laughed.

We had launched a cruise missile at Zarqawi a few days before—and missed. I figured that was what prompted Congress to junket over about twenty of its members to see how "our boys overseas" were faring. General Gorenc, I, and Brig. Gen. Allen Peck, deputy commander of the air war, escorted the delegation around Balad and showed them our Predator operations. The esteemed congressmen were on the bus, off the bus, like a bunch of seniors on a vacation tour to Branson. I gave them the standard spiel while they looked over *Ruby*, walked around her in the HAS, scratched their heads, and tried to look wise rather than puzzled.

"You mean, nobody *flies* this thing?"

"It's an RPA—a remotely piloted aircraft, sir. It has a pilot. It's just that the pilot is not actually *in* the airplane."

"Is that so? Sort of like a robot?"

"More like an aircraft that is remotely piloted," I retorted dryly, trying not to seem condescending.

They asked the usual questions. Does it have machine guns? How high can it fly? Not one person asked a reasonable question about general tactics, missions, support capability. I didn't think they were really very interested anyhow. They seemed to be more concerned about pounding their own chests. "Now, son, I want you to know that *I* voted for more exploratory funds to provide you boys with—" and "*I* helped get a bill passed that will—" That sort of thing. After all, they were members of the House of Representatives and always in reelection mode.

A few weeks later, General Gorenc asked me to accompany him to Baghdad's Green Zone to deliver an hour's general briefing on the Predator to another party of politicians, this time from the Senate. They were fresh off the plane and, it turned out, either more thoughtful or a little more weary than the congressmen.

Senator John Kerry from Massachusetts, chairman of the Senate Foreign Relations Committee, fell asleep, chin on his chest. Jet lag, I supposed. Or perhaps he hadn't yet recovered from the 2004 election campaign and his defeat by President George W. Bush. He asked no questions.

Senator John Warner was an old, craggy-faced character from Virginia, chairman of the Senate Armed Services Committee. He would have looked more at ease rocking on his front porch back home or windbagging at length on the Senate floor. This was his fifth term in the Senate, which made him the second-longest-serving senator in Virginia's history, behind only Senator Harry F. Byrd Jr. One of his distinctions was that he had once married actress Elizabeth Taylor. But then who hadn't?

Senator Ted Stevens from Alaska, chairman of the Senate Appropriations Committee, resembled a wizened little owl with horn-rimmed glasses—until he opened his mouth, at which time all traces of owl-like sagacity vanished. *Better to remain silent and be thought a fool than to open your mouth and remove all doubt.*

"Do you conduct combat search and rescue?" he asked.

A reasonable question.

"We do," I replied. "Say a helicopter goes down, we try to have a Predator first on the scene to coordinate the response, provide air support, and vector in a recovery team."

If he had just shut up after that.

"Do you ever think of landing the Predator to pick them up? You could sling a net under it, the guys could get in the net, and you could take off again."

I stared at him, not quite sure how to respond. I had just finished explaining how the Predator depended upon ground crews, LRE teams, and pilots.

"Well, sir," I drawled, trying to contain my sarcasm. "If we *could* manage to land a plane, the survivors would still be in the net when the choppers arrived. Then we'd have to figure out how to get the Predator out."

General Gorenc summed up the day in his inimitable manner. "Pogo," he said.

CHAPTER 25

Leaving Balad

THE SUCCESS OF BOBBY'S double air strike on the enemy mortar crews was due partly to a new P-Model version of the Hellfire missile. Previous M and K models had to "see" the laser reflection off the target as they launched, meaning that the pilot had to be flying directly at the target when he fired. The P Model, on the other hand, could be launched "off-bore sight"; Predator could fire the missile no matter which direction the aircraft was going. That saved precious minutes. After being launched, the missile turned and then acquired the laser energy for a high, low, or direct trajectory. High trajectory worked best for troops in the open, such as the mortar men. Once it was fired and acquired its target, the missile plunged more or less straight down on the mark to deliver an even distribution of shrapnel that wiped out everyone within a radius of the explosion.

There had been no further insurgent activity in either Area 200C or 200D since that night. That didn't mean the bad guys had quit. They simply shifted somewhere else, apparently thinking that whatever we used to nail them with before could not be utilized somewhere else. Silly rabbits. What Bobby and Falisha accomplished had to be duplicated in the other security areas around the air base in order to stop enemy infiltration. Balad had to be made secure since it was scheduled to remain active as one of the few superbases in Iraq when the United States drew down its force levels.

"Balad will be here, I believe, until the very end," General Gorenc predicted.

Area 200A posed our next challenge. We were receiving fire from that sector about every seven to ten days. Insurgents were like having a house infested with rats; the more of them you killed, it seemed, the more they bred. From HUMINT on the ground, a snitch, we learned that a former Iraqi colonel under Saddam Hussein was leading a small band of gunners and saboteurs in 200A. I made him a squadron priority. My pilots and I consumed hours in Watchdog Four-Six keeping an eye on his house or tracking his truck around the little city of Balad and the surrounding towns. He was bound to screw up sooner or later, and one of us would get a P-Model chance at him.

Kimberly Joan and I were flying Watchdog on the colonel's house late one afternoon when we noticed activity. Our man came out of his house in the company of a scrawny little bearded character in a *thawb* and turban, a stereotype that might well have escaped from the pages of *Mad* magazine. They got into the colonel's truck and headed in the direction of the air base and Area 200A.

At the outer edge of a field, they stopped, quickly unloaded a mortar tube concealed underneath a tarp in the back of the truck, and hid it in nearby shrubbery. According to ROEs modified under the new base security plan from my CONOPS, I was licensed to kill under narrow circumstances, which included hostile acts or hostile intent. However, before Joan and I could spin up a missile, the colonel and the Mad Bomber were back in their truck and gone. With the mortar tube hidden, "hostile acts" were over. To confirm hostile intent, I would have to wait for analysts relocated at the Joint Intelligence Center on the other side of the base to study the video feed and make a positive ID on the mortar.

I cannot confirm mortar, appeared in my chatroom as I followed the colonel's truck, hoping to get a shot. *Give me a few more minutes.*

I pounded my keyboard. *We saw it. It's a mortar.*
Stand by, Watchdog.

In the meantime, a U.S. Army patrol was conducting a raid on a house in a neighborhood on a spit of land surrounded on three sides by the serpentine Tigris River. The army called the area "Michigan" because of its shape. Our troops were in contact, taking fire from a flat-roofed row house, ducking and dodging from courtyard to courtyard as they used the cover of buildings and mud fences to advance on their objective. A tree was ablaze in the front yard of their target house.

Curiously enough, the colonel drove right down the street in Michigan while all this was going on. He had never gone there all the time we had been watching him. Instead of hauling ass out of the AO, as any normal jihadist would have done, he parked in front of a residence directly across the street from the contact location. He and his Mad Bomber cohort jumped out of the vehicle and ducked inside.

General Gorenc rushed into the GCS. I had notified him of the situation; he liked to be present whenever we fired a shot. I continued to circle Predator above the colonel's truck and the troops in contact while I explained what was going on.

"We're waiting for analysis confirmation," I said. "But we can't fire him up anyhow at his present location."

There were two reasons why I had to hold fire. First, it could be dangerous to nearby friendlies. Second, the crowded neighborhood made collateral damage possible. We didn't know who lived in the house or how many women and children might be present.

I was fast running out of time. I needed to land Watchdog in order to free up my GCS for launching the next mission for the Nellis POC. Staying on-station meant delaying support for some ground units that might need it. I contacted the liaison officer (LNO) in Qatar through the CAOC chatroom to request an in-flight extension.

Watchdog, request you provide a specific extension time came the terse reply.

We have a critical situation, I explained. *I need to fly until minimum fuel.*

Negative, Watchdog.

Fifteen more minutes?

Nothing but sullen, resentful silence in the chatroom. For certain rear-echelon types, daily priorities were set in stone and not to be juggled without specific instructions from one of the CAOC generals.

About two hours had elapsed since Kimberly and I picked up the colonel leaving his house to deposit the mortar in 200A. We could have dispatched a QRF to recover the weapon, but what we really wanted was the shooters. Best leave the mortar where it was—as bait. That would take time and patience since it might be a week or more before anyone showed up to use it.

We had enough evidence for a QRF to arrest the colonel and his Mad Bomber, if not to shoot them—as soon as analysts confirmed the weapon. In the meantime, all I could do was wait. And watch.

American troops on the ground busted into the house across the street, withdrawing a few minutes later empty-handed. It looked as though their guy had fled out the back door before I arrived overhead trailing the colonel. Soldiers climbed into their Bradleys, Strikers, and Humvees and roared off to some other hotspot. The tree in the front yard was still burning. Nobody came out to extinguish it.

As soon as the patrol left, the colonel and his buddy rushed out and sped off in the truck. I still couldn't figure out what they were up to. I followed.

Confirm mortar, Watchdog appeared in my chatroom.

The truck parked at a hardware store. The colonel went inside, came back out presently, then took the long way home. Every few minutes the LNO interrupted on the CAOC chatroom to demand that I pull off and land. I kept requesting extensions. Five times the LNO intruded.

General Gorenc was livid. "That goofy—. There's no reason

CAOC should be micromanaging us. We should make our own decisions on how to fly when we've got a hot target. We're in a better position to determine if our local mission will interfere with the schedule than they are in Qatar. I'm talking to General Peck about his providing better guidance to the CAOC floor."

Operation Neighborhood Watch had allowed my squadron a lot more autonomy. However, stove-piping remained an issue. Like all bureaucracies, the air force was full of it. There was often a disconnect between those who ran the war and those who fought it. Still, for all I knew, the stovepipe ran all the way to the White House, and CAOC was just an elbow in it.

I finally had to obey and pull off. The colonel and his Mad Bomber got away, at least for the time being. But I knew where he lived and I now had the goods on him. Even if we couldn't blow up his house, QRF had the green light to nab him. He and some of his buddies should be in Abu Ghraib prison by the end of the week. Or dead, if they returned to the mortar in 200A and we caught them there.

It was a strange cat-and-mouse game we played.

Senior Airman Kimberly Joan and I were on the flight line waiting to catch the medical evacuation flight to Ramstein Air Base in Germany, and from there back to the "real world." I had arrived at Balad with Captain Brent; I was leaving with Joan. My eyes swept across the HAS in front of which *Ruby* sat waiting for her next flight, and then across the sprawling air base simmering in the desert sun that had been my home for the past four months. I would miss Clark and the Okie; Statler, who had already redeployed; Falisha; and the other sensors and members of the squadron, even Brent with his kimchi breath and whining. Under stress, you grew close to your comrades.

Real life is not like the movies where all the strings were neatly tied up in the last reel, no matter how complicated things were in the beginning. I had been with Predator, and almost constantly at war, whether in-country or at Nellis, since March 2004. Returning stateside to be with Trish and help her through the rest of her treatment didn't mean the war was over for me. I would merely be trading LRE and line-of-sight Watchdog tactical mode for the larger operational picture of flying out of Nellis again. Neither Iraq nor Afghanistan was going to be a brisk ten-rounder after which the bell rang and everybody shook hands and went home. Both wars were a matter of endurance, like an old nineteenth-century bare knuckle fight that went on and on until one pugilist or the other was too bloody and beat up to continue.

I was departing with the feeling of having left too much undone. Although Operation Neighborhood Watch had curtailed much of the violence against the air base, insurgents continued to mortar Balad at intervals. Elsewhere, other things in-country seemed to have likewise changed little. Sunnis and Shiites with the cleric Muqtada al-Sadr and Abu Musab al-Zarqawi right in the middle of it all were infighting in Najaf, Nasiriyah, Fallujah, Sadr City, and Baghdad. Iraqi police recovered twenty bodies dumped into the Tigris River not far from "Michigan" where we had tailed the colonel that day. United States soldiers killed Abu Azzam, Zarqawi's aide, in a street fight in Baghdad; Zarqawi remained at large to declare another "all-out" war against the United States and Iraqi Shiites.

We still had not captured Osama bin Laden. Maybe I'd get another shot at him while flying out of Nellis. Saddam Hussein was on trial for crimes against humanity. We expected he would be found guilty and hanged. Couldn't happen to a nicer guy.

And the Pentagon announced that it was investigating allegations that U.S. soldiers posted photographs of dead Iraqis on a website in order to gain free access to Internet pornography.

"Honey, I've needed you," Trish e-mailed.

I needed her, too, to help ground me to a reality other than that of blowing up people seen on a monitor. Although so much was unfinished, I still felt I had contributed something to the history of aerial combat and made a permanent mark on the operational thinking of the U.S. military. At the modest change of command ceremony during which Maj. Sam Morgan relieved me as commander of the 46th Expeditionary Reconnaissance Squadron, operations commander Colonel Jacobs surprised me by commenting how "Major Martin has made the greatest difference of any of my squadron commanders."

A C-17, the medevac to Germany, taxied up while Kimberly and I waited. Ambulances arrived at the flight line to load that week's complement of broken, wounded, and frightened troops into the big bird's belly. They hobbled or were carried aboard on crutches and stretchers, wearing bandages underneath blue hospital pajamas, some even carrying their own IVs. One kid had both his legs blown off below his hips. Kimberly squeezed my arm in shock, and I heard her gasp. Neither of us could think of anything to say; we simply stared. This was closer to the war than we had ever gotten via monitors.

The kid with no legs smiled at Kimberly and me. "Sir . . . ma'am," he said in a Deep South drawl. "We're going home."

Kimberly sobbed quietly. She was like a mother to the whole world.

Witnessing all this helped me better understand Cindy Sheehan and her antiwar activism. You didn't have to agree with her or approve her tactics to understand her. Her son Casey was killed in Baghdad in April 2004. She had been on a crusade ever since then, beginning with the makeshift antiwar camp outside President Bush's Texas ranch. Last month, she and some two hundred other sit-in protesters besieged the White House, chanting slogans.

"The whole world is watching!"

"Mothers say no to the war!"

"Liar, Liar, Iraq's on fire!"

"War is terrorism with a bigger budget!"

Cindy Sheehan and the others refused to obey police officers' orders to leave and were arrested. From my distant perspective, I didn't think she would ever be emotionally whole again. She was a casualty of war.

All of us lost something in war, whether it be legs, arms, a sense of decency, or empathy for fellow human beings. I was a much harder man than I was two years ago. I kept going back to war because I knew that Predator and the crews flying it saved our soldiers' lives. Because of our work, not as many of our people would return home with missing body parts.

The professor in me, as well as the warrior, subscribed to the philosophy of John Stuart Mill. "War is an ugly thing," he wrote, "but not the ugliest of things. The decayed and degraded state of moral and patriotic feeling which thinks nothing is worth a war, is much worse. . . . A man who has nothing which he is willing to fight for, nothing which he cares more about than he does about his personal safety, is a miserable creature who has no chance of being free, unless made and kept so by better men than himself."

Even army specialist Lynndie England, I thought, might be considered a victim of war. While I was on my way home, she was on trial at Fort Bragg and probably destined for prison. She and eleven others of the 372nd Military Police Company were about to be convicted of torture and prisoner abuse that occurred at Abu Ghraib prison in Baghdad. Photos of her posing with naked prisoners and leading them around on a dog leash had been distributed around the world. Another strategic failure. One didn't have to approve of particular conduct in order to understand it.

Each of us reacted differently to being at war. In retrospect, perhaps I should have recognized symptoms in Captain Brent that might have been called "shell shock" during World War II and Korea. I had nursed him along when the better approach might have been to send him home, no matter the damage to his career. He had been scheduled to redeploy with Kimberly and me, but his relief hadn't yet arrived. He had been moping around for days, stuffing himself with kimchi and mourning over how he missed his wife and kids.

Two weeks after I left Balad, as I was to learn, he launched an airplane for a Nellis pilot and then got up and walked out of his

GCS before he handed it over, leaving the Predator to fend for itself. My relief, Maj. Sam Morgan, sent him to the flight surgeon to have him checked out emotionally and psychologically. On that same day, the hospital happened to be full of maimed and dying Iraqis from the attack of a suicide bomber that had just occurred in downtown Balad. All those crying children and wailing mothers. Brent freaked out. Sam sent him home straightaway, even though that left him short a pilot and meant canceling missions until Brent's replacement arrived.

Later, Captain Brent testified before a flight evaluation board (FEB) that he was certain he had properly transferred control of the Predator; the computer log proved otherwise. I testified in his behalf. However, the FEB reviewed the evidence and ruled that he should be taken off the line, never to fly again, although he could keep his wings. Since he was only a few years away from retirement, he was offered a desk job until he could complete his service and separate with honor.

From Ramstein and Frankfurt, Kimberly and I caught the rotator to Baltimore, then on to Las Vegas. I walked off the jet way at Las Vegas' McCarran Airport in civilian clothing with a bag slung over my shoulder. I stopped and looked eagerly across the faces until I spotted a crop of strawberry roan hair that could belong to only one woman. Trish. My Miss Ruby.

PART IV

Nellis Air Force Base, Nevada

CHAPTER 26

On the Hunt

I TOOK A COUPLE of weeks off in Nevada with Trish, just enough time to put my home life back together again. By that time, Predator and its mission were becoming known to the general public through the media—*mis-known*, rather, to use a President Bush euphemism. What people seemed to hear was that we were a bunch of bloodthirsty, video game–playing, missile-slinging cowboys who indiscriminately shot up a town and then went out for a beer. Fighting like that was so *unfair*.

Some people would look at me strangely. "Let me get this right," they might say, and I knew what was coming. "You're out there on the air force base killing innocent people on the other side of the world while they can't shoot back at you?"

I tried to contain my temper, I truly tried. Not always successfully. Sometimes I broke down my response to a few words. "You have no idea what you're talking about."

Back in my old GCS at Nellis, I often found myself patrolling along the Iraqi-Syrian border looking for infiltrators. During the Soviet incursion into Afghanistan, foreigners from all over the Muslim world rushed in to see if they could get a piece of the Russians. It was a mark of status for a jihadist to boast how he had fought Russkies. Status from Afghanistan was what thrust Osama bin Laden into his leadership role with al Qaeda.

It was becoming an even greater status symbol for students of Arab colleges and madrasas to brag about killing infidels from the United States. Michael Ware of *Time* magazine, one of only a few journalists crazy enough to venture out of the Green Zone in Baghdad, wrote a piece describing how the insurgency that began as an Iraqi movement was being taken over by foreigners much better trained and equipped to fight a proper guerrilla war. Abu Musab al-Zarqawi, for example, was a Jordanian. Others were Saudis, Syrians, Yemenis, and other Arabs who slipped into the country to organize themselves along the Afghan Mujahidin model. The Syrian border was a sieve through which would-be martyrs and mad bombers made their way to Baghdad, Najaf, Fallujah, Mosul, Kirkuk, and Basrah.

While at Balad in Iraq, most of my flying was conducted in support of local security and local forces. At Nellis, I returned to dealing with the global picture of terrorism. I began flying 80 percent of my time in support of SpecOps—Delta Force, SEALs, and Green Berets, as well as other government agencies wandering around the area of responsibility (AOR). The war was a lot simpler in-country than it was 7,500 miles away from it where I not only had day-to-day challenges to contend with but also the broader picture of operational tasking.

Army and marine general purpose units were assigned tactical areas of responsibility in which they were accountable for keeping the peace, managing reconstruction projects, assisting the new government, and developing indigenous forces to some level of competence whereby one day they might take over their own country. Certain special forces, on the other hand, were apparently free to do pretty much whatever they wanted, ranging theater-wide to collect intelligence and launch unannounced raids. Their job, as I understood it, was to work the borders and hunt down and capture or kill most-wanted insurgents and terrorist leaders, such as Zarqawi and Osama bin Laden.

These groups mostly worked independently of the normal military command structure and were so secretive that I sometimes thought they didn't even talk to one another. Even though Predator crews were cleared top secret and therefore authorized to be briefed on every aspect of an operation, "Op-4" communicated with us cryptically, if at all. I understood to some degree why they were so cautious about sharing information. It was, after all, *their* ship that would be sunk by loose lips. Still, they could be a pain.

One morning, one of these secret outfits gave me the grid to a house in Husabayah, directly on the Syrian border, separated from it only by a wire fence. I was asked to "take a look" at it. Okay. I took a look. Typical Iraqi house, mud colored, flat roofed, low mud-brick fence around a courtyard, rising sun sparking through a couple of date palms in the back yard.

What am I looking for? I asked through the chatroom.

Suspicious activity.

So what do you want me to do?

Watch it.

After a while, as I circled, three Blackhawk UH-60 helicopters appeared on my screen, hauling coal low and just inside the border fence. They drew rein like cavalry above the house. Someone from the bay of one of the choppers let fly a propelled grenade, which blew in the front door. The choppers dived down like hawks after a rabbit and then kissed the streets fore and aft to let out about twenty-five battle-geared troops, who promptly stormed the house.

You can leave now. Thanks, Speck. That was my call sign for the day.

Anytime.

I had no idea what it was all about; I wasn't "read in." Obviously, that was the way these guys wanted it, all hush-hush. They often came in like gangbusters and then left everything in a mess for local troops to clean up. Free-ranging behavior like that tended to foster resentment from regular grunts and their commanders.

To be fair, they sometimes scored big-time in capturing wanted subjects. Nonetheless, it seemed to make little difference in the overall scheme of things. New leaders trickled in. Iraq was as violent as ever. Our guys were still being killed or going home without their legs. If Iraq was any nearer democracy and self-sufficiency, I contended, it was due to conventional army and marines, not SpecOps. If Predator took the 80 percent of our time currently directed to special forces and turned it to tactical support of patrols and convoys that were IED'd and getting shot up every day, we could save a lot more lives and make more noticeable progress.

Special Operations was on the hunt. Bradleys and Humvees rumbled up a street toward a house in Husabayah, on the Syrian border. For weeks the little town had been a hotbed of activity for foreigners infiltrating into Iraq. My task was to hang out overhead and watch for anyone who tried to squirt out the back of the house when the raid began. We called them "squirters." As usual, I didn't expect the special forces guys to tell me what it was all about, who they were after, or why they might want him dead.

It was midmorning. Bright and sunny, the way it was most days of the year. Pedestrians stopped to stare. A few chased after the procession in order to watch the action, the way people run to a fire or a fatal traffic accident on the freeway.

Troops encountered resistance the moment they turned the corner from down the block. Whoever was in the house must have seen them coming. Muzzle flashes sparked from windows. Tagalong spectators split for safer parts, lifting up their man-dresses and hauling ass with their bare legs flashing, looking back over their shoulders as though the Twelfth Imam was about to return in a chariot of fire pulled by a mule.

A guy bolted out the back door of the house. He wore a checkered headdress and baggy pants; few of the foreigners who came to fight wore robes unless they were going to a formal event—for example, a meeting with Zarqawi on how to bomb Shiites on a bus or in a mosque.

The guy in the checkered headdress ran down a cluttered back street and jumped over a courtyard wall to hide in a neighbor's tin shed. I got on the radio, reporting in the abbreviated language we used when things got hot. "I have a squirter, out the back door." I might have been in Nevada, but when something like this happened I was right there.

"*Nitrate Three-Five, keep watching him,*" came the response from the ground. "*We're a little busy. We'll get to him.*"

Busy they were; it looked like a pretty good firefight. By now all the Iraqis who came to watch might have encountered a little more excitement than they had bargained for. They were hiding behind houses or sprawled face down in gutters with their arms over their heads. Next time they wouldn't be so eager to tag along.

My squirter was more nervous than all of them put together, as he had a right to be. Like the proverbial cat in a room full of mousetraps. He would have been really wired had he realized I was watching him.

A couple of grenades went off down the street. The guy stuck his head around the open side of the shed, jumped back inside, then poked his head out again. Apparently deciding this wasn't the safest place for the average jihadist, he vacated the shed like a cat with its tail on fire. He skidded into the narrow gap between the nearest house and a high mud wall, where he went to ground again. I lost sight of him temporarily, but he couldn't go anywhere without my seeing him. Ever patient, I kept watching.

About ten minutes passed. Things were winding down back at the house. Soldiers were escorting out insurgents with their hands flexicuffed. One was either wounded or dead. The soldiers dragged him out by his arms.

Below me, a Blackhawk helicopter joined the fray.

"*Nitrate, do you have eyes on?*"

"Roger that. See the house with the satellite dish on the roof? He's down between the house and the fence."

"*Light him up. We're going to shoot him.*"

"Roger . . . Hold on, he's running again."

The guy vaulted the wall, sprinted across the open adjacent yard, and disappeared into a thicket of small trees directly behind a house. He didn't come out the other side. I could have made short work of him with a Hellfire, but SpecOps often relegated Predator to observer status.

"Movement in the trees," I reported, and beaded him with my infrared marker.

The Blackhawk came in hot over the trees and yard. A fiery blast ripped out the back of the house and left the trees standing.

"Missed!" I snapped, hoping my disgust carried. "He's still there."

The next rocket took out the middle of the yard and left the trees standing.

"He's still there," I taunted. "You sure you have my sparkle?"

Suggesting the SpecOps guys might not have their act together was tantamount to throwing down the gauntlet. Exasperated and embarrassed, the pilot brought the gunship around a third time. The squirter was probably too scared to move. This time the bird let loose with everything it had, hosing down the yard, the house, and the trees with 20mm machine-gun fire and a few rounds of 40mm automatic grenades for good measure. Overkill. Fire and smoke roiled up and obscured the entire neighborhood.

This time they got him and got him good. The trees were no longer standing.

"*Thanks, Nitrate.*" I could almost hear the sneer in his voice.

A few days later, I was back on the border with SpecOps spying on a couple of houses separated by a mud-brick wall that is so ubiquitous in these parts. Insurgents kept jumping back and forth across the wall from one house to the other. I had no idea what was going on. All the unit on the ground told me was, "We're trying to get those houses bombed."

If fast movers were coming in, I required further details in order to avoid a possible midair collision. Flight path, direction of approach, altitude, weapon types, run-in heading, airspace control authority.

On my chatroom screen appeared *Classified. Need to know only*.

Madness. No way could I stand by and depend on the supported ground unit to keep me out of the way of approaching fighters. I went around him and contacted a marine combat controller.

Know anything about fighters cleared to bomb a target in my area? I asked through the chatroom.

Speck One-Five, two marine fighters are cleared at ten thousand.

Roger that. Request descent and AO update.

I noticed in my other chatroom that Carriage from CAOC was weighing in and insisting upon an immediate air strike against the two targets, which meant they were high priority. As commander of the air war in Iraq, Carriage had authority to order strikes without coordinating with the local ground commander. I switched over to the Warhawk radio frequency, the air support coordination net for all of Iraq, and found everything in total confusion. The discussion seemed to have been going on for some time. There appeared to be some discrepancy between coordinates relayed to the strike aircraft and the actual targets. I listened in for a moment or two.

"*Warhawk, this is Sparrow One-One. Target description does not match the coordinates.*"

Warhawk's role was to relay instructions and communications back and forth between the CAOC and the tasked aircraft.

"*Carriage, this is Warhawk. Sparrow cannot match target with the grid.*"

Sparrow refused to conduct the mission until discrepancies could be resolved. The point would have been rendered mute were the fighters using GPS-guided bombs. They could simply have plugged in the coordinates, and the smart bombs would have been on their way.

Laser-guided munitions, however, were something different. Pilots first had to identify the targets, eyes-on, correspond them to the grid coordinates, and then designate them with a laser to guide the bombs to impact. Confusion somewhere in the chain could result in blowing up the wrong houses, killing innocent people, and letting the bad guys escape.

I could easily identify the houses as I had been watching them for well over an hour. From all the way back in Nevada, I designated the targets with my Predator's laser, keyed my mike, and guided the Sparrows to their targets. Minutes later, both houses exploded in fireballs. Whoever the guys were down there wouldn't be jumping back and forth across any more walls.

I couldn't help myself. Say the Devil made me do it. I pounded on my keys and sent the message: *Are we still "need to know only?"*

CHAPTER 27

Hanging Saddam

AMERICAN TROOPS HAD PULLED Saddam Hussein out of a hole in the ground on December 13, 2003, nine months after the war began. Dirty, unkempt, and cowering, he appeared more animal than human after so long buried in the earth. Since then, the man around whom three major wars were fought—the Iraq-Iran War in 1980, Desert Storm in 1991, and Iraqi Freedom beginning in 2003—had been held in the custody of U.S. forces at Camp Cropper in Baghdad waiting for a proper Iraqi government to try him for "crimes against humanity." For three years his case remained unresolved as the war raged on and on and the death toll rose. I flew missions over Baghdad, looked down upon where he was confined, and puzzled over the depravity of human nature that spawned sadistic men like him and allowed them to attain such power.

History revealed that humanity has allowed totalitarian rule and submitted to it from almost the first day man got onto his hind feet and built fire. The twentieth century had been the world's most destructive era—and it appeared that the twenty-first century would follow suit. Two great wars had been fought, in addition to countless revolutions, civil wars, coups, and genocides—a century of almost constant strife and discontent. While tyrants may not have created the conditions, they certainly rode them with bit, spur, and crop.

Saddam Hussein was arguably among the world's top ten despots when it came to ruthlessness. He fit well in the company of such dictators as Adolf Hitler, Josef Stalin, Fidel Castro, Pol Pot, Mao Tse-tung, Idi Amin, Benito Mussolini, Robert Mugabe. All believed, or at least professed to believe, in an ideology of a socialist utopia that justified their suppressing, torturing, and slaughtering their own people in the name of creating a perfect world. In fact, they were largely evil cynics who cared more for personal power than the lives of their country's citizens. They started wars and were the cause of them. They amassed great personal wealth while keeping their countries in poverty, typically destroying their own nations in the process.

All of them had so far ended badly. Hitler shot himself while hiding in a bunker; a mob hanged Mussolini; Idi Amin died in exile; the Khmer Rouge arrested Pol Pot and he died after being tortured; Stalin expired a paranoid and miserable old man under rumors that he may have been poisoned. I expected Saddam Hussein to meet a similar fate. The drama of his trial dominated my tours in Predator, beginning with his first hearing on July 2, 2004.

At this hearing, Saddam looked clean and well fed and possessed restored confidence, not at all like the scruffy character that American troops pulled out of his hole eight months before. He called the court a "ploy" designed to help President George W. Bush win reelection.

"This is all theater," he raged. "The real criminal is Bush. I am still the president of the Republic and the occupation cannot take that away."

He concluded by defending his 1990 invasion of Kuwait, referring to the Kuwaitis as "dogs" who were trying to turn the women of Iraq into "two-penny whores."

Human Rights Watch, Amnesty International, and various United Nations bodies pointed out that the Iraqi Special Tribunal failed to meet international standards for a fair trial. United Nations Secretary General Kofi Anan declined to support the proceedings, while Ramsey Clark, former U.S. attorney general, volunteered for Saddam's defense team. Of course none of this changed the fact that Saddam was suspected of killing up to three hundred thousand of his own people.

And the fact that he had gassed thousands of Kurds and orchestrated the murders of forty of his own relatives was not in dispute. Nor was the fact that alleged prostitutes were beheaded and political dissidents taken into the desert, shot, and buried in mass graves.

On the one hand, I thought it would be a gesture of goodwill by the Iraqi central government in the eyes of the world to meet the standards of international justice. And what better way to prove that justice was re-established in Iraq than by treating Saddam to the same due process as we believed all Iraqi citizens deserved. On the other hand, I suspected that the typical Iraqi just wanted to get the whole thing over with so they could move on with their lives.

The trial for Saddam and seven of his former henchmen began shortly after I returned from Balad and my second Predator tour to Iraq. Saddam wore a dark suit with a white shirt and a tie while his codefendants appeared in traditional Arab male dress. They were all seated in a row in a pen constructed of white metal bars, facing the Iraqi Special Tribunal of five judges.

"I do not respond to this so-called court," Saddam sneered.

Judges asked for his full name, for the record. He refused to provide it, saying, "You know me. If you're an Iraqi, you know me. I am the president of the Republic of Iraq. I did not say 'deposed.' "

The trial proceeded week after week in a similar disruptive vein. I often flew a mission and got home in time to see Saddam on CNN. Always he was arrogant and demanding and ready to defy the world. He looked out at the judges from his cage, telling them to "go to hell," insisting he would not return to the courtroom. After two lawyers from his codefendants' team were kidnapped and murdered, he went on a hunger strike to protest lack of international protection.

He missed one meal.

The trial would drag on for more than a year. In the meantime, at least thirty thousand Iraqis were killed or injured by insurgents in the aftermath of the dictator's rule. Saddam himself assumed the role of victim, accusing his American captors of having tortured him.

"I have been beaten on every place of my body, and the signs are all over my body."

He refused to be examined.

On November 5, 2006, the Iraqi Tribunal would finally find him and six of his codefendants guilty of genocide on several counts, including the gassing to death of thousands of Kurds in Halabja in 1988 and the slaughter of thousands of Shiites during their uprising in 1991. Sentenced to hang, he roared, "Long live the people! Long live the Arab nation! Down with the spies! *Allahu Akbar!*"

He was to be hanged just before dawn on December 30, 2006. A video of the execution showed him wearing a black overcoat as three masked guards led him into a room inside the heavily fortified U.S. Green Zone. As the noose tightened around his neck, one of the executioners shouted, "Long live Muqtada al-Sadr!"

Defiant to the end, Saddam, who had refused to wear a hood, uttered his final words with a grunt of contempt and in a mocking tone, "Huh! Muqtada al-Sadr."

Iraqi judge Munir Haddad, a witness to the hanging, said Saddam appeared "totally oblivious of what was going on around him. I was very surprised. He was not afraid to die."

Another witness, however, thought he was a "broken man. He was afraid. You could see fear in his face."

Death came "in a blink of the eye after the executioner activated the gallows just after six a.m."

Iraqi national security advisor Mowaffak al-Rubaie announced the death to the world. "This dark page has been turned over," he said. "Saddam is gone. Today Iraq is an Iraq for all the Iraqis, and all the Iraqis are looking forward. The Hussein era has gone forever."

Iraqi Americans celebrated in the streets of Dearborn, Michigan, home of the nation's largest concentration of Iraqis. And in Iraq, there were also street celebrations throughout the day—accompanied by deadly car bombings against Shiites in Baghdad and Kufa. Because of the difference in time zones, it was getting late when I heard the news and saw the video on TV. Trish and I went to bed; she was feeling considerably better following aggressive steroid treatment against her lupus. I had to get up early to fly a mission the next day.

The War on Terror continued.

"Sir, may I see your driver's license? The reason I stopped you is because you were speeding through the last intersection."

The Las Vegas cop was young with expressionless features and an earnest, no-nonsense voice. My mind hadn't been on driving until I heard the blip of his siren and saw blue lights flashing in my rearview mirror. I was in a hurry to get to the POC to relieve the off-going pilot on time. What was the matter with this policeman? Didn't he realize I was at war? At any other point in history, it would have been inconceivable that a combat pilot could take time out from fighting to have a leisurely at-home breakfast with his wife and then get a speeding ticket on the way to work. Another of those strange juxtapositions of alternate lives from two vastly different worlds.

An hour later I was in the middle of a firefight in Al Asad. It was late afternoon in Iraq. Marines were taking fire from a two-story flat-roofed house in the suburbs. Everything on the fires net, the frequency for coordinating close air support and artillery, was going ape. Guys were screaming at one another over the radio while commanders tried to restore order, all against a steady punctuation in the background of rifle and machine-gun fire.

"*Break! Break! Break! Oscar One-Six, tell me what you got there.*"

"*We're taking fire, Delta X-Ray. Small arms from a house. Estimate several personnel . . . oh, shit!*"

"*What's your grid location, Oscar?*"

"*How the fuck . . . ? Taking fire. Stand by, Delta, for GPS.*"

"*This is Delta X-Ray. Go with your SITREP* [situation report], *Oscar. All other units, clear the net. Say again, clear the net.*"

"*Delta, this is Oscar. I'm popping smoke.*"

The pilot of a marine Cobra attack helicopter came up on the air. "*I spot smoke. Acknowledge green?*"

"*This is Oscar. Confirm green.*"

A thin finger of smoke in that color eddied up from in front of the house. The house was fairly isolated from its neighbors with a large courtyard signifying an owner of some means. I circled above and watched marines hurling grenades from behind the courtyard wall.

The explosions flashed bright on my screen when they detonated, followed by more small-arms fire banging on the radio.

The Cobra gunner, Avenger Four-Three, was a qualified forward air controller—airborne (FAC-A), meaning he could clear me to take a shot. I checked in with him on the radio to let him know I was ready to support.

"*Shade Four-Five, we have two or more insurgents shooting at marines. Take a look and see what you can see. How copy?*"

"Shade Four-Five, good copy." *Shade* was today's call sign.

From past experience, I expected the shooters to scoot as soon as things got too hot for them. Sure enough, as marines maneuvered on the house, two figures sprinted out the back and scurried down a narrow, cluttered alley as though their butts were afire. Both carried rifles, probably AK-47s, the weapon of choice in that part of the world. Russia and the Chinese must have manufactured millions of them.

I reported it and tailed the two via my cameras until they darted into a large grove of palms and I lost sight of them. My crew and I circled to keep an eye on them, but they failed to reappear. Everything began to settle down on the radio frequency. Marines, now unopposed, busted into the house, ransacked it, and came up empty-handed. They returned to their vehicles and, at my direction, headed toward the palms.

"*Shade, this is Avenger Four-Three*."

"Go for Shade."

The chopper made loops a few thousand feet below me. "*Shade, we think the insurgents returned to the house after we left and are hiding there.*"

"Avenger, I haven't seen anybody run back out of the trees."

Avenger Four-Three insisted it was good intelligence, that our insurgents had somehow eluded me and returned to the house. Seemed a stupid move to me.

"*Shade, we need you to shoot a missile into the house. Kill 'em or run 'em out. How copy?*"

Maybe he was right. He was a lot closer to the action than I was. "Shade copies. Ready for nine-line."

The FAC-A passed the required information to make sure we agreed on the target and the effects to be produced. Then he followed up with some additional instructions.

"Shade, time is 1503 Zulu. Time-on-target 1508 so we can clear troops out of the area. Call when inbound for clearance on final."

"Roger that, Avenger. Four minutes to inbound."

"Roger. Your run-in heading is two-seven-zero so we can stay clear. Continue."

Senior Airman Abado, my sensor, and I spun up the P Model and armed it. I flew a wide run and calculated when to turn inbound for a precise time-on-target (TOT). The house filled my screen. The green smoke had dissipated.

Suddenly, Avenger broke air, shouting, *"Shade, abort! Abort! Say again, abort!"*

I saw them at the same time. A couple of Humvees pulled up in the street in front of the house. Somebody hadn't gotten the word. It was 1506 Zulu. I veered off.

Minutes later, the Humvees drove off in a hurry. Avenger cleared me again for a time-on-target of 1511 Zulu. I turned inbound and was cleared hot at 1509:50.

I triggered my left missile at 1510:33. At 1510:59, a perfect TOT, the missile slammed straight down through the center of the roof. My screen flashed and pixilated. When it recovered, I saw smoke billowing out the hole in the roof and fire flapping from broken windows.

No one ran out of the house.

"Let's take another spin at the palm grove," I suggested to Airman Abado.

Marines had scanned the grove, but I wasn't sure they had done so thoroughly. And I still wasn't convinced that our mice were stupid enough to return to a mousetrap.

This time, we spotted movement down in the palm shadows. The insurgents must have ducked into a tunnel to hide and were now venturing out after the marines left. They were on their bellies crawling around and craning their necks, trying to see without being seen. I could have easily killed them with my remaining Papa missile, except Avenger had relinquished tactical control to a ground JTAC. Ground opted to clear an F-18 with a GBU-12 laser-guided 500-pound bomb.

I concentrated my camera on the target and waited in anticipation for the *ka-boom!* Instead, all I got was a puff of dust from the middle of the trees. The bomb was a dud.

The insurgents shifted about uneasily. Instead of running while they had the chance, however, they foolishly decided to stay put. The F-18 made a second pass, this time with a Maverick missile. No dud, it made a very nice *ka-boom!* and an even nicer mushroom. It took out one end of the grove, splintering palms, igniting fires, and cratering a hole about twenty feet across.

By now, the insurgents were probably getting the idea that someone knew where they were hiding. The survivor rabbited out from the smoke, abandoning his dead or wounded comrade. Every jihadist for himself.

I called the chase from the air on the JTAC frequency. No grass was growing under the guy's feet. He paused only long enough to bury his AK-47 in the back yard of a nearby house, getting down on all fours to dig a shallow cache with his hands, sending dirt flying like an eager dog after a gopher.

"He's moving again. South," I reported.

Soon he arrived at another house. I assumed it to be his, judging from the reception he received. About a dozen kids and women ran out and mobbed the guy, obviously glad to see him again this side of Paradise, even though a bit the worse for wear. They couldn't have helped hearing the gunfire and explosions and likely knew what their hero had been up to.

They all went inside quickly, shut the door, and stayed there. Satisfied that he had gone to hole, I directed the marines to him. My crew and I were no longer needed. We wouldn't be killing any women and children just to take out one insurgent foot soldier. I pulled away for my next tasking. At the end of my shift, I would hand the airplane over to the next crew, file my after-action report, note the missile fired in my logbook, and go home to have dinner with Trish.

And try not to get another speeding ticket.

CHAPTER 28

Under the Whole Sky

SEEN FROM THE AIR, the Golden Dome Mosque of Samarra, on the east bank of the Tigris River north of Baghdad, reflected sunlight like a polished penny, the dome having been covered with seventy-two thousand gold pieces and surrounded by walls of light blue tiles. Also known as the Tomb of the Two Imams, referring to Imams Ali and Hasan, it was one of Islam's holiest sites, exceeded in veneration only by the shrines at Karbala and at Najaf, where al-Sadr's Mahdi Militia had holed up.

This dominant feature of the Samarra skyline never failed to impress me. Today I circled it twice from an altitude of seven thousand feet, zooming in my cameras in order to detect any unusual activity. Since the beginning of the war, more than 425 foreigners and several times that many Iraqis had been taken hostage by Islamists, among them 31 foreign journalists. Victims were sometimes secreted inside a mosque; insurgents knew that mosques were generally off-limits to U.S. troops.

It seemed I was always searching for signs of the latest kidnap victim. Jill Carroll of the *Christian Science Monitor* claimed the latest dubious honor of having been seized by insurgents. A new Crips gang in town, the so-called "Brigades of Vengeance," was threatening to kill her unless their demands were met. Although only five of the thirty-one kidnapped journalists had been murdered, none of them female, her life was nonetheless in peril. All the various gangs, most of whom owed loose allegiance to Zarqawi and al Qaeda of Iraq, had proved their propensity for public violence and cruelty.

Carroll had become an international cause célèbre since her abduction in Baghdad on January 7, 2006, partly, I suspected, because she was young (twenty-eight years old), attractive, and an American. People were reacting sort of like movie audiences did when the ape in *King Kong* took the girl. Every American in Iraq, including Predator crews at Balad and at Nellis, were on full alert to find and rescue her before the jihadists executed her and dumped her headless body in an alley with a note pinned to it saying something like "It's the will of Allah."

She had been covering the war in Iraq since October 2003 for the *Monitor*. On the day she disappeared, she had motored to the Adel district of Baghdad with her driver/interpreter Alan Enwiyah to keep an interview appointment with a prominent Sunni politician named Adnan al-Dulaimi. She arrived at his office to find him gone.

Disappointed, she and her escort left. They made it about three hundred yards from Dulaimi's office before several masked gunmen jumped in front of the car, pushed Enwiyah over from behind the wheel, and sped off with both the journalist and her interpreter. Enwiyah was later found nearby shot to death. Jill seemed to vanish, except for the occasional missive delivered by her captors to either Al-Jazeera TV or one of the other Arab stations that catered to insurgents. Each time that happened, our search for her intensified with new life and hope.

At a news conference nearly two weeks after the abduction, politician Adnan al-Dulaimi declared how "this act has hurt me and makes me sad. We are against violence by any group, and we call on the government and U.S. forces to stop raiding houses, arresting women. I call upon the kidnappers to immediately release this reporter who came here to cover Iraq's news and defend our rights."

I might have been suspicious of his posturing had I been investigating the case and not simply flying around overhead looking for clues or sign as to her whereabouts. For all I knew, he may have set up her kidnapping, considering how he had made an appointment with her that he failed to keep.

On January 27, Al-Jazeera aired a twenty-second video that showed Carroll standing in front of a plain white background. The Brigades of Vengeance, a narrator explained, threatened to kill her unless the United States released all female prisoners currently being held in Iraq. Trouble was, the United States was holding few if any females.

Three days later, a second video found its way to Al-Jazeera. This one, without audio, showed her crying while wearing a traditional Arab headdress called a hijab.

On February 9, after Carroll had been in terrorist hands for over a month, a third video appeared in which she was allowed to speak. She wore full Islamic dress and sat in a chair in front of a large floral pattern.

"Today is Thursday, February 2, 2006," she said in what was apparently a statement supplied by her captors. "I'm with the Mujahidin. I'm here. I'm fine. Please, just do whatever they want, give them whatever they want as quickly as possible. There is a very short time. Please do it fast. That's all."

A Kuwaiti TV station, Al Rai, reported the next day that Carroll's abductors had communicated a deadline of February 6 for their demands to be met—or the young woman reporter would be executed in Allah's name. People "close to the kidnappers" allegedly told Al Rai that she was "in a safe house owned by one of the kidnappers in downtown Baghdad," living with a group of other women.

The Brigades of Vengeance, whoever or whatever it was, knew how to keep a secret. Hours of flying to scan mosques, houses, and apartment buildings turned up nothing. Neither had our raids on the ground against suspected sites. Our intel came up empty. As far as we knew, Jill Carroll might already be dead. Zarqawi and others of his ilk certainly weren't above beheading a woman, although they probably wouldn't do it for the edification and entertainment of Arab webcast viewers. We would simply find her body one morning floating in the river.

The more I learned about our enemy, the more I viewed Islam extremism as a self-centered and perverted religion. There were many places in the Arab world where women were little more than chattel, like goats or sheep. That was quite obvious in the way women were treated. During winter rains, men and boys rode in the sheltered cabs of trucks and pickups while their wives, mothers, and sisters huddled together in the open truck beds. In the Taliban's Afghanistan, a Muslim man of forty could legally marry a six-year-old girl and consummate the marriage when she turned nine. After all, the prophet Muhammad himself did likewise when he was forty-seven years old. A nine-year-old girl in Nevada was in the fourth grade.

It was mind-boggling trying to understand a religious interpretation that condoned honor killings; whipped or stoned women for transgressions of "morality"; practiced male polygamy; extolled martyrdom in Allah's name; rarely condemned suicide bombers who blew up innocent men, women, and children in public places; publicly beheaded its prisoners while chanting "Allahu Akbar"; murdered cartoonists who dared use Muhammad's name in vain; and placed fatwas on the heads of errant journalists and writers.

In her book *Infidel*, Ayaan Hirsi Ali, a Somalian raised in the Muslim world, recorded how she escaped to Holland and experienced a kind of "awakening."

"[Islam] spreads a culture that is brutal, bigoted, fixated on controlling women, and harsh in war," she wrote. "[M]ost people *think* that Islam is about peace. It is from these people, honest and kind, that the fallacy has arisen that Islam is peaceful and tolerant. But I could no longer avoid seeing the totalitarianism, the pure moral framework that is Islam. It regulates every detail of life and subjugates free will. True Islam, as a rigid belief system and a moral framework, leads to cruelty. . . . Their world is divided between 'us' and 'them'—if you don't accept Islam you should perish."

While I had nothing personal against Islam, I could see her point. And I held little hope for Jill Carroll's safe recovery.

Nothing appeared to be going on around the Golden Dome. The SpecOps unit that I supported asked me to check out another mosque in Baghdad on a tip that "activities related to the kidnapping" might be going on inside. I and other Predator jocks had been running back and forth like that for the past three months.

On February 23 I flew over Samarra and was jolted to discover that the Golden Dome Mosque was no longer a landmark. The shining penny was gone, obliterated the previous morning at daybreak when several men belonging to Zarqawi's al Qaeda group forced their way into the shrine, tied up the guards, and set off two bombs. The explosions destroyed the mosque, leaving only the minarets still standing and relatively undamaged. The minarets would likewise be blown up the following June.

Since Sunni jihadists hated Shiites worse than they hated Americans, acts such as blowing up the Shiite Golden Dome was presumably intended to precipitate a civil war between Sunni and Shiite. Whether the bombing had anything to do with Jill Carroll's abduction was an unanswered question. I was savvy enough by this time to understand that virtually everything the jihadists did in Baghdad was somehow related. Carroll might be merely another pawn in the whole mess, part of Zarqawi's plan to win Iraq by first destroying Iraq, riding chaos and spilling blood all the way.

Smarter people than I were trying to make sense of it. Other than finding Jill Carroll, my strongest urge at the moment was to locate Abu Musab al-Zarqawi and place a Hellfire squarely on his forehead.

Allahu Akbar!

During World War I, doughboys spent months living and fighting in muddy, moldy trenches—static warfare at its most savage and elemental level. Korea was pretty much more of the same. Grunts fighting the Vietnam War slogged endlessly through rice paddies and jungles seeking sporadic engagements with the enemy. Now, in the first decade of the twenty-first century, nothing better illustrated the changing nature of modern warfare than the Predator. We who flew RPAs were literally ushering in a new age that was global in nature and anything but static. One day we might be soaring high above Iraq tailing bad guys, seeking kidnap victims, blasting mortar teams, or keeping an eye peeled for al Qaeda leaders. The next day—or even that same afternoon—we could be prowling the Afghanistan border with Pakistan seeking bad guys of a slightly different stripe, perhaps, but nonetheless possessed of a similar ruthless and fanatic nature.

The difference between future war and past war was dramatically impressed upon me one afternoon in the rugged mountains along the Afghan-Pakistani border, not far from the famous Khyber Pass. A narrow trail zigzagged through snow-splotched escarpments as it traversed the sides of sheer-rock cliffs and hung precariously over dizzying canyons. The Nellis crew of one of our Afghan Predators, call sign *Skybird,* spotted four men hiking the trail, each carrying a weapon and a pack of impossible size on his scrawny back, each as agile as a mountain goat. The soldiers of Alexander the Great would have identified with them. So would the Mongols of Genghis Khan. Wars had been fought like this for thousands of years.

But wars for others had changed remarkably in ways that the four men down there could hardly comprehend. It never occurred to them to look up except to search for helicopters and other easy-to-spot warplanes. That an eye from the inner edge of space might be watching was too far-fetched for them to imagine.

Today's *Skybird* crew was British working at Nellis with Americans to gain experience in combat RPA flying. The pilot, Flight Lieutenant Kevin Gambold, gave me a heads-up on the situation through his chatroom. I was in the POC as mission commander, a position that I, one of the more experienced pilots in the squadron, was assigned more and more. I switched my attention to *Skybird*'s monitors, where a message in the chatroom appeared from the army Special Ops task force on the ground in Afghanistan.

Skybird, they're hostiles, read the communiqué from the commander of Operation Achilles. *We had a scrap with the enemy and several slipped away. These guys are them. I want them dead.*

In Afghanistan, all types of military, nonmilitary, or quasi-military actors populated the stage: U.S. Army, U.S. Marines, special forces, CIA, Taliban, al Qaeda, opium smugglers along the old Silk Road. JTACs on the ground frequently had no access to video feed, which meant they had to rely on verbal information relayed through sources nearer the action. As always, the ground commander had the final say. A target was fair game as long as he could demonstrate a hostile act or hostile intent, and no friendly forces were in the vicinity.

Skybird, I've checked with everybody, the Achilles commander continued. *We have no friendlies in the general area.*

He then ordered an air strike. *Type III control, cleared to engage.*

It was not our place to question him. He was there, we weren't. Nonetheless, Flight Lieutenant Gambold wavered. The War on Terror was very unpopular in Britain, with its large and increasingly rambunctious Muslim immigrant population. The British government imposed such strict rules of engagement on its war fighters that, in most cases, the English could shoot only in self-defense.

As far as I was concerned, those four men down there had painted bull's-eyes on their own butts by shooting at our guys. I dispatched an American crew to take over *Skybird*'s controls and settle the refined British conscience. We Americans weren't quite so finicky over what we shot at. How could you have a war if you weren't allowed to shoot the enemy when you got the chance?

The four Taliban were by now walking single file through a mountain pass. Two were close together in the lead while the other pair trailed behind. Not bothering to move tactically, they seemed to be out for a stroll, obviously thinking themselves safe in their lofty redoubts. Wait until they got a load of what we had in store for them.

Skybird, with its new pilot, initiated an attack run, turning inbound from its circular orbit. This particular mission carried only one missile, so the plan was to take out the two leaders then track the squirters to wherever they fled and call in other assets to finish them off.

Skybird fired its missile. The unexpected happened during Hellfire's thirty-second flight. The two leaders halted, plopped down on the ground, and leaned back on their packs. It was break time. The other two caught up and joined them in a little circle, all totally unaware of doom already released and screaming toward them out of the sky. It didn't get any better than this. Four birds with one stone.

The missile impacted directly in the middle of the powwow and went high order. The picture on my MCC screen washed out from the explosion. It quickly cleared again to show the bright rise of the fireball.

Once the smoke and fire dissipated, I saw four mangled and scorched bodies blown back onto the ground, the contents of their packs strewn all over the landscape. The four Taliban were predictably dead.

War wasn't fair, particularly in the new age of warfare. Anyone who thought it was hadn't been there.

How could you figure it? Just when everyone was about to give up on finding Jill Carroll alive, she turned up at the Sunni Islamic Party headBobbys in western Baghdad and handed the clerk a letter in Farsi from her kidnappers. It asked that she be helped.

She told the press that she had been freed unharmed and had been treated humanely over the period of her captivity. The only condition of her release, she said, was that she first agree to tape a video in which she criticized the occupation of Iraq and praised insurgents as "good people fighting an honorable fight." The video promptly appeared on the same jihadists websites that previously carried tapes of beheadings and attacks on American forces.

I honestly found no fault in her behavior. She was a twenty-eight-year-old woman who for three months had been completely under the control of her captors. What she ate, when she ate, where she slept, what she wore, when she could speak, almost what she could think were all at the whim of her abductors. A lot of civilians would have done the same thing, *anything* to stay alive. Her reaction upon returning to her home state of Michigan tore at my heart and made me realize how much we Americans took our lives for granted.

"I really feel like I am alive again," she said. "I feel so good. To be able to step outside anytime, to feel the sun directly on your face, to see the whole sky. These are luxuries you just don't appreciate every day."

Jill Carroll did not return to Iraq. She became a city firefighter.

"We're glad she's alive," President Bush said.

CHAPTER 29

Until the Job Is Done

THE UNITED STATES ISSUED the first patent for a wireless telephone in 1908, although a hand-held practical mobile phone other than in a vehicle did not appear until 1973. Mobile phone networks were starting to become common by the 1980s. Nearly five billion cell phones were in worldwide use during the first decade of the twenty-first century.

A study by Motorola found that one in every ten phone subscribers kept a second secret phone for activities such as extramarital affairs and illicit business transactions. Drug dealers, organized crime, gangbangers, child predators, and terrorists all depended on cell phones. What many of them failed to realize was how easily, barring legal niceties, these phones could be invaded by law enforcement and intelligence services.

For terrorists, especially, the cellular telephone proved to be both a boom and a curse. Satellite technology allowed the U.S. National Security Agency, the CIA, and other clandestine agencies to listen in on conversations between suspected terrorists anywhere in the world. It was almost impossible for a terrorist to hide with his phone. Using a technology known as multilateration, the geographical location of a cell phone could be closely determined by just its roaming signal, whether or not a call was being made.

Insurgency leaders in both Iraq and Afghanistan understood that electronic signals could be traced by various means. That made them careful to speak in code and metaphor and to switch phones often. However, they often became careless over time.

When I was an ROTC cadet at Purdue, we ran in formation to the cadence of an old marching chant from the Vietnam War entitled "Napalm Sticks to Kids."

> *Flying down a river bed, one hundred feet high;*
> *Drop that napalm, watch 'em fry.*
> *Kids in the schoolhouse trying to learn;*
> *Drop that napalm, watch 'em burn.*

A new generation of tech-savvy warriors might change the chant to:

> *Flying over Baghdad at ten thousand feet;*
> *Listening to those beeps and squeaks.*
> *Hajji in his house talking on the cell,*
> *Hellfire's coming, send him straight to hell.*

The Combined Joint Special Operations Task Force in Afghanistan traced the cell phone signal of an important Taliban leader to a baked-mud hut on the outskirts of Kabul and requested that the POC at Nellis take a look with a Predator. As mission commander, I sent one of our planes. We now had two of them operating in-country.

The target house sat in the middle of a fenced-in compound in a semirural community with only distant neighbors. Our cameras picked up women and children outside busy washing clothes in zinc-and-wooden tubs, sweeping off the doorsteps, and carrying water. Special Operations said the guy who interested them, the Taliban leader, was inside, undoubtedly plotting his next nefarious crime. We couldn't get him without killing his family, and we couldn't ring him on his cell and say, "Hey, Mustafa, how's it hangin'? Why don't you get your ass out the door and away from the kids so we can shoot it off?"

What we could do was wait. The airplane was full of gas, and we were patient. Plus, Predator in Afghanistan was now packing iron.

A couple of hours passed while we orbited out of sight and downwind to muffle the slight hum of the engine. Two men eventually exited the back of the house and meandered into a nearby field, apparently seeking better reception or privacy from the wives and kids in order to talk terrorist business. The in-country JTAC lit up the mission chatroom.

That's him. We're getting SIGINT [signals intelligence] *now.*

Our man was transmitting. Target confirmed. I instructed my pilot to spin up a missile and plan his attack run. The JTAC passed a nine-line and asked my pilot to contact him when he was inbound.

Approach from the west. Cleared hot.

It was a perfect setup, a perfect shot. I experienced the by-now-familiar pixilation on the screen as the missile launched from its rail to briefly interrupt the return link. The two men were still standing within a few feet of each other thirty seconds later when the Papa streaked straight down to impact between them. They never knew what hit them. The smoke cleared and I observed two charred corpses lying in the field next to the blast crater.

Sometimes it didn't make a lot of sense, fighting the Global War by killing one or two guys at a time. Still, I supposed if we kept at it long enough and killed enough of them, the violence would eventually subside to the point that our rebuilding efforts could gain traction and we'd be on our way to bringing some sort of stability, if not prosperity, to the country.

In the meantime, on to the next SIGINT hit. *We're listening. Can you hear us now?*

The Combined Air and Space Operations Center (CAOC) in Qatar headed up all joint air operations in that theater of the world, which included Iraq and Afghanistan. Within the CAOC in order of rank appeared the Coalition Forces air component commander (CFACC—a three-star general), his deputy (a two-star), the CAOC director (a full-bird colonel), and the chief of combat operations (CCO—a lieutenant colonel), who oversaw the day-to-day execution of the air war. A senior intelligence duty officer (SIDO) and a senior offensive operations officer (SODO) worked for him. Because the Predator was first of all an intelligence asset, RPA operations fell underneath the SIDO.

Further down the pecking order—and among the team of company-grade intelligence officers who monitored Predator missions for the SIDO—came those who manned the intelligence, surveillance, and reconnaissance cell (ISARC) at some dingy rear-echelon desk in the back of the CAOC inside a top-secret vault. The ISARC's primary function was to build daily Predator schedules and incorporate changes when needed. As mission commander at the Nellis AFB POC, I soon fell out of favor with the ISARC and thus the SIDO, due to the way I viewed my mission responsibilities, a stand that threatened my career.

Too often, senior people in the air force considered both the MQ-1 Predator and its big brother, the MQ-9 Reaper, as assets that did little more than orbit over a target for hours on end. It was often difficult for these people to imagine that our flying an aircraft from the other side of the planet required the integrative skills and judgment of a combat aviator that far exceeded those of the vernacular Xbox teenager. In fact, RPAs in general and the MQ-1 in particular were routinely called upon to conduct complex missions such as close air support and armed overwatch. At the time of writing, the Predator has logged in excess of seven hundred thousand combat hours in CENTCOM (Central Command) and delivered more than a thousand precision weapons in danger-close scenarios that required aviators trained in tactics, techniques, and procedures.

The main factor that led to the blurring of lines between intelligence gathering and traditional strike operations lay in the nature of counterinsurgency and counterterrorism operations. Counterinsurgency targets were hard to find and easy to lose. A target that might take days to locate and identify could disappear within minutes. That meant that standalone weapons systems like Predator must have the capability to conduct operations anywhere along the find-fix-track-target-engage-assess kill chain. A recent army colonel covered it best: "The OIF [Operation Iraqi Freedom] fight has morphed into a conflict where all sensors are shooters and all shooters are sensors."

My concept of Predator operations as mission commander and that of a shavetail ISARC lieutenant at the CAOC in Qatar came to a head when some of our SpecOps people found themselves up against a superior Taliban outfit along the Afghan border. One of my planes riveted its attention on the platoon-sized Taliban element as it sneaked through the hills above the SpecOps team in an attempt to cut it off. My live-action screen in the POC showed what the Predator saw; a map screen displayed an icon of a B-1 bomber maneuvering toward the engagement to drop eggs in the Taliban nest.

The enemy platoon took cover in a cave when the bomber began dropping ordnance. Using targeting radar—a less than precise technique—the B-1 blew up the entire landscape without getting near the cave. Not even close. Its last munitions fell a good half-kilometer off-target. Eventually, the Taliban would figure things out and hightail it. Another lost target.

I directed the pilot of the nearby Predator to get on the radio with the B-1 pilot. "Iron Man, this is Rover Two-Six. We have eyes and can provide support to correct fire."

"Much obliged, Rover Two-Six. Standing by."

My Predator pilot and Iron Man went into communications while I monitored. Things were proceeding nicely toward a resolution when the ISARC lieutenant at Qatar came up in my chatroom.

POC-N, have your airplane RTB [return to base]. *Its intel tasking is complete.*

The guy didn't seem to know what was going on, although he should have. Under other circumstances, or had the Predator's next mission been a critical one, I would never have questioned him.

However, the only reason for the airplane to RTB was to stay on a schedule. That wasn't good enough. After all, the pilot assured me he had plenty of fuel.

Annoyed, I typed back, *We're in the middle of a weapons engagement. I've told my pilot to remain on-station.*

Take your plane off-station, POC-N.

I didn't relish a turf fight, but we weren't going to desert our troops down there when they needed us.

The airplane stays on-station, I typed back, *until the ops mission is complete or until bingo fuel.*

The lieutenant threw a chatroom tantrum. His next response came back in all caps, to express his displeasure. *MOVE AIRPLANE NOW.*

Negative.

My phone rang. It was the chief of combat operations at Qatar, a lieutenant colonel. He sounded pissed off.

"What the hell is going on?" he demanded.

I explained and dug in, saying, "Sir, I'm keeping the plane on-target until we finish the job. Five minutes tops."

"You sure you want to do that?"

"Yes, sir. I am."

He hung up.

The boot ISARC lieutenant in the shadows at CAOC wasn't about to give up. After all, I had challenged his authority. To hell with reason and common sense. He continued to blast me with all caps, insisting my plane move. Losing my temper, I blasted him back.

THAT AIRPLANE STAYS ON-STATION UNTIL THE JOB IS DONE. YOU CAN TELL THE DEPUTY CFACC, WHOM I KNOW PERSONALLY, THAT MAJOR MARTIN IS THE ONE WHO MADE THE CALL.

With Predator's assistance, the B-1 pounded the cave, sealing the Taliban fighters inside forever. I instructed my Predator crew to RTB as directed, advising the pilot that if he pushed up the power a bit, we could still keep on the ISARC's schedule.

I should have known that wouldn't end the spat. It didn't matter whether or not I was right. The offended lieutenant sent the chat excerpt to his boss, the SIDO. *Look at this guy. He's disobeying orders.*

Colonel Chris "Sponge" Plamp, my squadron commander, summoned me to his office the next day.

"You did the right thing, Major," he agreed, "but you shouldn't be telling them who you are. If you've got an issue, tell them to call me. I provide top cover."

I hesitated to guarantee him it wouldn't happen again, even though I knew that my career couldn't survive another such confrontation. The way I saw it, my primary job for Predator was to provide support for our troops in the field, try to save as many of them as I could. If the shoe clerks in the CAOC didn't understand that, well, I could make a few suggestions as to what they could do to themselves.

"Major, I know you're a warrior trying to do the right thing," Colonel Plamp said. "But you need to keep your emotions out of it."

I resolved that I would at least not volunteer my name should another showdown occur.

Which it did. A few days later, the shoe-clerk argument took up where it left off, again precipitated by a situation in Afghanistan similar to the first. Some of our people were under heavy fire. The on-ground JTAC requested that Predator remain on-station to direct in other assets and to provide cover if needed. The ISARC at CAOC said it was time to move on.

Unable to do so, I replied through the chatroom. *We need to stay and provide coverage.*

We have a schedule, POC-N. You're violating the rules.

I tried to be reasonable. *The next target is low priority. These guys are troops in-contact. We're needed here.*

The electronic shouting began. *THIS IS AN ORDER.*

I followed Colonel Plamp's advise and held my temper. *Stand by. We'll need a few minutes to coordinate airspace.*

It was an obvious ploy to delay. Nonetheless, it bought us enough time to wrap up the situation on the ground and usher our guys out of harm's way.

Lieutenant Colonel Plamp promptly received another call from the CAOC. This time it was the full-bird colonel in charge of operations. "Who was that guy?" he demanded.

Colonel Plamp hedged in an attempt to provide top cover as promised.

"Colonel Plamp, you tell me who that man is."

He did so. He had little choice.

"Uh-huh. Major Martin. I knew it."

Although I had yet to receive orders, my next rotation was scheduled for the CAOC, where I would be working as part of the liaison team under the full-bird colonel's direct authority. Talk about tossing meat to the lions.

"Sponge, you send that guy to me. I'm going to kick his ass."

"Colonel, sir. Give him a chance and he'll be a good man for you."

"Well, you send him out here. I'll show you how we treat guys like him."

The next day, I was back in Sponge's office.

"Send me, sir," was my reaction. "I'm ready to go back. If he'll give me a chance, I'll show him I can do a good job."

"I can't do that to you, Matt. I fear it's become a personal vendetta out there. We can't let that affect the mission."

I would not be returning to the desert after all, at least not right away. In order to try to cool things down and not let the incident influence my career, Colonel Plamp demoted me off mission commander rotation and sent me back to the line, flying in the squadron like other pilots. I felt like I had been kicked in the guts. It seemed that principle didn't matter. What mattered was playing the game. Trish tried to console me.

"It'll all work out for the best, Matt."

She was always my little optimist, the plumb line that kept me straight.

CHAPTER 30

Back in Sadr City

THE PHANTOM THAT WAS Abu Musab al-Zarqawi had haunted me for the past three years, from my first predator combat tour over Iraq and the fight at Fallujah. I was obsessed with the fear that he would escape justice after what he had done to Nicholas Berg, the kidnapped contractors, and dozens of others. During those years, I expended hours and days flying over Baghdad or Fallujah scouring streets and buildings for some sign of his presence, staking out safe houses, providing air cover for raids that always turned out to be dry holes, hoping that one day I would get a shot at him. Sometimes I thought I knew as much about this man, my nemesis, as I knew about my wife.

Wily Zarqawi had been fighting the war a lot longer than I. Seventeen years ago, in 1989, he left Jordan to befriend Osama bin Laden and hitch up with the insurgency in Afghanistan against the Soviets. By that time, however, the war was all but over and the Russians were pulling out in defeat. Disappointed and still looking for action against the infidels and a chance to build a new Arab empire devoted to Allah, Zarqawi ended up back in Jordan doing a nickel in prison for attempting to blow up the Radisson Hotel in Amman, part of a conspiracy to overthrow the monarchy and establish an Islamic caliphate.

After Zarqawi's release from prison in 1999, he traveled to Afghanistan and joined Osama bin Laden in establishing a militant training camp specializing in poisons and explosives near Herat and the Iranian border. After the 9-11 attack on America, he moved his operations to Iraq and set up sleeper cells in several cities to resist the U.S. occupation. Bin Laden stayed behind in Afghanistan.

From 2003 until the present, Zarqawi has been implicated in thousands of killings that ranged from beheadings to suicide bombings. While escaping capture time after time, he continued to taunt President Bush and the West in videos released through Al-Jazeera. The latest video, in April 2006, showed him sitting on a floor with an AK-47 resting across one knee while he addressed a gang of black-hooded men.

"Your mujahidin sons were able to confront the most ferocious of crusader campaigns as a Muslim state," he praised his radicalized supporters. He then focused his ire on the United States and spoke directly to President Bush, revealing no awareness of irony regarding his statement in the context of the Islamic martyr complex. "Why don't you [Bush] tell people that your soldiers are committing suicide, taking drugs and hallucination pills to help them sleep. By Allah, your dreams will be defeated by our blood and by our bodies. What is coming is even worse."

Zarqawi's "Tawid and Jihad," known for attacks on Shiites and videotaped beheadings of foreign hostages, had merged with bin Laden's al Qaeda in 2004 to become known as al Qaeda in Iraq. By 2006, Zarqawi was the most wanted man in Jordan and Iraq. The U.S. government offered a $25 million reward for his capture, the same price it had placed on the head of Osama bin Laden.

Zarqawi's three wives, the second of whom was fourteen when he married her, pleaded with him to leave Iraq temporarily and issue orders from outside the country. One of the wives recalled his response later when interviewed by a journalist.

"He gave me an angry look and said, 'Me? Me? I can't betray my religion and get out of Iraq. In the name of Allah, I will not leave Iraq until victory or martyrdom.'"

As it turned out, that day was to come sooner than he expected. And the Predator played a role in it.

Everything came together against him on the late afternoon of June 7, 2006, about six months before Saddam Hussein was hanged. For the past thirty days, my squadron had burned over two thousand hours of Predator flight time keeping a sharp eye on several safe houses in and around Baghdad known to be frequented by Zarqawi. The most likely of these was a remote residence about two miles northwest of Baghdad near the community of Baqubah. Other RPA pilots and I expended at least six hundred hours flying over it the month prior, watching and waiting for the terrorist leader to show up.

It was a small baked-mud farmhouse surrounded by a few palm trees and several outbuildings. A rural road ran in front of it. When I was flying, I alerted every time a car pulled up to the house. My crew and I zoomed in our cameras, hoping to identify this most valuable target in Iraq when he finally made an appearance. Nothing would have satisfied me more in my air force career than to be involved in taking down the mad butcher of Fallujah. Lamentably, it was not to be.

Unknown to me at the time, our intel people in Iraq had traced a senior al Qaeda figure to Jordan, where he was arrested. Acting on information gleaned from him and from a variety of other sources, intel learned that Zarqawi was going to meet with his spiritual advisor, Sheikh Abd al-Rahmann, at the Baqubah home late in the afternoon of June 7. Unaware that anything different was about to happen, I got off a recon mission over that same house and went off-duty. I could have kicked myself later for not hanging around.

United States special forces tracked al-Rahmann to the rendezvous. Zarqawi arrived shortly thereafter and was positively identified. A few minutes after six o'clock that evening, local time, two air force F-16 warplanes dropped a pair of 500-pound bombs on the house, reducing it to rubble. Zarqawi was still alive when on-ground SpecOps reached him, but he died shortly thereafter in an ambulance on the way to a hospital. Also killed were his spiritual advisor, al-Rahmann, and four others, including Zarqawi's second wife and their child. It was all over in six minutes, rather anticlimactic in light of all the effort the United States had put into finding and killing the terrorist leader.

This one was for Nicholas Berg and all the others the monster had beheaded.

Back when Dad and I were hay farming in Indiana and involved with horse people and racetracks, an anecdote went around that illustrated how hardheaded and stubborn some folks could be. According to the parable, a racehorse owner heard of a mule in Arkansas that could outrun every Bobby horse and Thoroughbred in the state. Anticipating winning some big money, the racehorse owner bought the mule and entered it in the First at Blue Ribbon Downs. The starter gun went off, the gates blew open and—the mule refused to leave the gate.

Feeling put upon, the mule's angry new owner loaded up the animal and hauled it back to Arkansas to demand his money back.

"There's one thing I forgot to mention," the previous owner explained. "That mule is one stubborn son-of-a-buck. Take a sledgehammer with you to the next race. Just before you put him in the gate, hit him right between the eyes as hard as you can with the hammer. That'll get his attention."

Trish always said I was stubborn like that mule. I preferred to call myself resolute, tenacious. Having been keelhauled from mission commander back to the line crew over the dustup with the CAOC's ISARC was like being hit between the eyes with a hammer. It caught my attention. Not that I would have handled anything differently. I felt I was right, and when I was right . . . well, maybe I *was* stubborn.

That same stubborn streak compelled me to strive to excel at whatever I undertook. Call it the midwestern work ethic. Lieutenant Colonel Plamp assigned me an additional task in my downtime that eventually led Predator flying into new realms and possibilities and revitalized my career.

The numbers of Predators and other RPAs like the MQ-9 Reaper were not only outpacing the growth of conventional recon aircraft on the war fronts, they were also increasing faster than new pilots could be trained to fly them. In an attempt to make better use of available pilots, the air force designed a prototype GCS called a multiaircraft control (MAC), which allowed a single pilot to fly more than one airplane at the same time. The concept was simple. It was like taking several basketballs and spinning each one in turn on the gym floor until you had all of them spinning together.

One MAC pilot could fly up to four aircraft from a single control station. He put an aircraft into the sky and assigned it to recon a set pattern, say in orbit around Kabul or Najaf. Then he picked up the second plane, the third, the fourth, assigning each to its own little chunk of airspace. Monitoring all four screens, he could take physical control of any one of the planes if the situation required it—if, for example, he needed to tail an insurgent leader in Baghdad, or if a couple of jihadists planting an IED in Nasiriyah were begging to collect their virgins.

Everything went along fine as long as three of the four aircraft were only punching holes in the sky. Trouble began whenever two situations arose at the same time, such as two targets to prosecute or two malfunctioning airplanes. That was more than one pilot could handle. Although MAC was therefore a flawed concept, it did give us an idea.

In a country as big and barren as Afghanistan, and with aircraft as slow as Predator, it often took a long time to fly from one point to another. With a sensor operator assigned to each plane to take over copilot duties like fuel checks and airspace coordination, one pilot could easily monitor several transit aircraft under MAC. The planes were at least smart enough to fly a programmed route. Trying to actually work all these airplanes alone when they arrived at their destination was where the pilot needed help.

I therefore modified the MAC concept through a new system called monitored transit operations (MTO). I set up tests, configured hardware, wrote procedures, coordinated routes, and trained crews. One pilot handled a MAC transit, but when the planes reached their destinations and were working, four GCSs and four crews took over to fly five sorties at the same time—a 20 percent increase in crew utilization that allowed us to provide more coverage with no additional resources. It also eliminated the problem of a pilot being confronted with more tasks than he could cope with.

The leadership at Nellis was so pleased with the results of my MTO testing that it moved me to the top of the list to fill an opening on HeadBobbys staff. Being good at something generally meant you were promoted from whatever practical task you were performing to supervising, teaching, or a staff position. In the new job, I would serve as chief of the Predator/Reaper Operations Branch at Air Combat Command HeadBobbys, Langley Air Force Base, Virginia. My mission was to organize, train, and equip forces for the MQ-1 Predator and the newly developed MQ-9 for deployment to combat commands.

I doubted if pushing papers would be as action packed as pushing plastic airplanes—but at least I had weathered the CAOC stain, and my career was back on track.

On January 20, 2007, President George W. Bush infused another 30,000 troops into the nearly 140,000 already in Iraq and named Gen. David Petraeus, U.S. Army, to command the Multi-National Force in Iraq and oversee the new counterinsurgency "strategery" that became known as "The Surge." According to Army Field Manual 3-24, primarily developed by Petraeus, the key to suppressing the insurgency lay in a philosophy of "clear-hold-build."

Rather than remaining forted up in large bases and undertaking grand sweeps, small ground units would push into communities to establish security stations and combat outposts that allowed them to live and fight among the people. "Clear" the area of insurgents, "hold" against their return, and then "build" through political and community reconstruction. This required battalions to independently plan and execute their own operational missions, each battalion commander in essence becoming its own little joint force commander, each in charge of his own assets.

This countered air force doctrine that had been in effect since the Battle of Kasserine Pass in 1943. In Tunisia, Allied airpower was assigned within boundaries in support of individual units rather than utilizing its ability to shift and cross lines to support higher priorities. This distribution of airpower into so-called "penny packets" ended in catastrophe for Allied forces—the U.S. Army's II Corps was routed and pushed back fifty miles with heavy casualties.

After "The Surge" began, the U.S. Air Force returned to decentralized operations for the first time since 1943. By this time we had twenty-four combat air patrols (CAPs), the secretary of defense having pushed us to field as many Predators as possible. The patrols were allocated directly to division commanders, who further reallocated them down to brigade and battalion level. Baghdad, for example, had two dedicated Predator missions.

The "clear" phase of "clear-hold-build" worked somewhat differently in larger cities than in villages and the countryside. A company or a battalion could simply sweep through a town of a few thousand people and kill, capture, or run off local insurgents, who had few places to hide once they were flushed. In Samarra and Baqubah, as in Fallujah and other large cities, U.S. troops surrounded the city and removed noncombatants while keeping insurgents contained. Then they stormed in to "clean up."

From my staff position at Langley, I traveled periodically to Creech AFB (where the 15th Reconnaissance Squadron had been relocated) to keep flight current in the Predator. When I returned to Nevada in 2008, I found decentralized operations working much better than I expected, largely due to the expanded number of Predators now distributed throughout the war zone. Crews knew their areas, their targets, and the personalities behind the chat. More than at any other point in the war so far, they were closer to "fused" operations, in which intel and ops were conducted simultaneously. It was very much like we had practiced in Balad with my line-of-sight program. And we were flying so many CAPs that all priorities were covered (something that wasn't true in 1943), avoiding the prioritization problems of decentralized control through sheer numbers.

I also discovered some unfinished business in Sadr City. Cleric Muqtada al-Sadr, "ole Chubby Cheeks," remained at the center of Shiite resistance, using the slums as a safe haven from which to lob rockets and mortar rounds into Baghdad's Green Zone. Finally, the generals had had enough of him. They ordered a wall built along the edge Sadr City, a ten-feet-tall, two-feet-thick fence of concrete barriers to isolate the suburb from the rest of Baghdad. A precursor to a major new offensive like the one that had cleared out Fallujah. Al-Sadr would finally get his comeuppance.

It had been nearly four years since I took my first Hellfire shot at a target in Sadr City. Since then, I had lost count of how many weapons engagements I had participated in as either shooter, instructor-mentor, or support. Almost as soon as I stepped off the plane in Las Vegas, I was back in combat in Sadr City.

The wall around Sadr City was almost completed. United States soldiers were beginning to drop leaflets and broadcast messages on radio and TV warning people that they must leave for their own safety through a number of designated American checkpoints. Realizing what was about to happen, insurgents were attempting to slow down the wall's completion through the use of snipers to harass construction workers. My mission for the day was to protect construction crews from shooters.

Viewed from above, the slums of Sadr City resembled a maze of rat nests, buildings all jumbled and mismatched, with trash, old construction materials, and junk cars mounted on blocks littering the streets. I was patrolling the wall from 10,500 feet, about two miles up, when the JTAC in a Humvee on the ground came up on the radio.

Barnes 4-6, this is Horseman 3-2. One of the construction crews is taking fire.

My crew and I shifted our focus to the grid he supplied. All we saw were workers and a few soldiers crouching behind the wall. The insurgents had gotten a little smarter since the days I flew regularly in Iraq. They had figured out that if they took potshots from the open, albeit from behind some barrier, we could somehow spot them from the air—even if they couldn't see the plane watching them. Therefore, they started building "duck blinds" of plywood or canvas on the flat roofs of buildings. While the blinds offered little cover, they were somewhat effective at blocking off at least some of the infrared radiation from the insurgents' bodies to make it more difficult for us to find them with IR sensors.

By triangulating the direction of hostile rifle fire, I homed in on a makeshift plywood shelter on the third-story roof of a house overlooking the wall and about a hundred meters inside it. The shelter resembled a large kitchen table. My IR detected heat signatures of several people moving around inside and underneath the blind. I transferred the image to the JTAC in his Hummer.

Barnes, it's the sort of thing we're looking for. My battle captain is prepared to declare the site hostile. Can you go lower for a better look?

Roger that, Horseman.

Rules of engagement permitted an air strike during a hostile troops-in-contact. I spun up a Papa missile as I circled wide to avoid other air traffic while I dropped altitude to 6,500 feet. In the meantime, the ground battle captain authorized me to kill the sniper position.

I started my run from west to east to shoot away from the wall. That way, if the missile failed to track or the laser malfunctioned, which sometimes happened, the Hellfire wouldn't hit our own troops. If it impacted in the tightly packed residential district, it was better than pounding our own guys.

Barnes 4-6 is in from the west.

Barnes 4-6, you're cleared hot.

I turned the aircraft inbound at eight klicks out, and then placed the target on the nose for a shot at 5.1 klicks out. My sensor operator cross-haired the target. I depressed the launch enable button on the throttle and pickled the trigger on my control stick. Fourteen seconds later, I saw the missile streak in from the top of my screen and slam dead center into the tablelike structure.

The damage I inflicted became clear once the picture reemerged on my screen. Most of the house's third floor had collapsed into the second floor. Flames flapped. A water pipe had burst, shooting an arc of water into the street and soon flooding it. Some people down the block hid underneath a pushcart. I later learned that a team of army troopers found what was left of three men in the wreckage on the roof, along with an RPG launcher and several rifles.

Barnes, a perfect shot. Good work.

A good shot in a long war.

I circled the sP in my logbook after the shift ended, rushed to the ops center to write my AAR, and then rushed some more to McCarran Airport in Las Vegas to catch the next airliner back to Virginia. On the way, I pulled my rental car into an In-and-Out Burger, changed out of uniform, and grabbed a burger and fries. Only after I was standing in line to board my flight did I have time to reflect on what had just happened, how unreal this war truly was.

I flew nearly 3,000 miles to climb into a stationary cockpit and fly an unmanned warplane 7,500 miles away to find some angry poor people and kill them. Then I caught a commercial air carrier to go 3,000 miles back home to have breakfast with my wife. I still hadn't discussed with Trish the details of what I did. But one day, maybe.

Epilogue

THROUGHOUT HISTORY, EACH INNOVATION in warfare has been accompanied by claims that it would make war "important and obsolete" (to quote Ronald Reagan). The bow and arrow was so terrible, it was said, that it would end warfare. The same claims in turn were made about firearms, warplanes, machine guns, tanks, bombs, missiles. Each advancement succeeded in allowing warriors to kill other warriors with more accuracy and with greater and greater distance between them and peril. Instead of these innovations ending warfare, however, a former secretary of defense feared that "we may be lowering and dropping the bar to war" as we depend more and more upon robots to do our fighting, and as the public understands that the risks of war are mainly being assumed by American "machines."

The Iraqi War was the first in history in which viewers could download video of combat from the Web in the security of their own homes. Nearly ten thousand video clips of combat footage were downloaded off YouTube alone in 2008, much of it captured by flying unmanned aircraft. "War porn" fans have become the equivalent of sports fans, a sort of entertainment block.

It makes little difference whether optimism or cynicism accompanies the continued advancement and use of unmanned combat systems. Science and technology in the field, as in all other fields, would become increasingly and inevitably more sophisticated. I tried to be optimistic in viewing innovation in war machines as steps toward making America safer and more prosperous in a world filled with danger. I also believed from my own experience that machines would make warfare less destructive, not more, and that it would become less and less feasible for America's enemies to challenge us—driving our adversaries to the bargaining table rather than the battlefield. This has certainly been the case for both the existential nuclear threat of the former Soviet Union as well as major conventional warfare.

Either way, sci-fi had entered the real world.

In 2009, Defense Secretary Robert Gates emphasized the vital role that RPAs and UASs, had assumed in the War on Terror since 9-11. The numbers of unmanned aircraft had increased twenty-fivefold over the past eight years to a total of more than five thousand. As of January 2010, the air force had more than eight hundred active-duty pilots flying the MQ-1 Predator, the MQ-9 Reaper, or the high-flying RQ-4 Global Hawk. The demand for pilots is expected to reach at least 1,400 by 2013, with equal numbers of sensor operators and mission intelligence coordinators.

"There are Predators airborne in support of Operations Enduring Freedom and Iraqi Freedom twenty-four hours a day, three hundred sixty-five days a year," Col. Christopher Chambliss, 432nd Wing commander, pointed out. "Because of what they can bring to the fight, the Predator is the most requested asset in-theater."

The MQ-1 had so far flown more than seven hundred thousand combat hours in Bosnia, Kosovo, Iraq, and Afghanistan, among other countries, and employed over a thousand weapons in danger-close scenarios with a valid kill rate of over 90 percent, leading to a new type of irregular warfare previously thought impossible in urban environments. As an evolution of the MQ-1, the MQ-9 Reaper is twice as fast, flies twice as high, and carries fifteen times the payload. Nearly half of all new aircraft to be purchased through the last presidential budget proposal would be unmanned. If that trend continued, it wouldn't be long before 30 to 50 percent of U.S. Air Force warplanes would consist of stealthy, high-speed, long-distance, large-payload unmanned aircraft.

Some of those would be optionally either manned or unmanned, depending on combat and tactical requirements. A cockpit could be rolled onto an aircraft for a manned mission during phase I of a major contingency. Once air superiority was achieved, the cockpit could be rolled off and replaced with fuel for longer-duration unmanned sorties. Aircraft crews would need the skills and be prepared either to fly in the air or fly the planes from a ground GCS.

New generations of RPAs were already taking to the skies to keep an eye on enemies and potential enemies and sting them if necessary. The latest model of unmanned stealth reconnaissance aircraft, the RQ-170, also known as "the Beast of Kandahar," resembled the much larger swept-wing B-2 Stealth bomber. Although the military confirmed in 2010 that this aircraft was being used in Afghanistan, specifics as to its capabilities remained highly classified.

In March 2009 the air force, working with Pentagon researchers, approved the development of a prototype of the next U.S. spy craft—a giant unmanned dirigible that could remain aloft for as long as ten years while flying at an altitude of sixty-five thousand feet, safe from enemy fighter planes and out of range of most missiles. Known as ISIS, for integrated sensor in the structure, the aircraft has a giant airborne radar system that would be capable of providing ground operators with close details of vehicles, planes, and even people hundreds of miles away.

Increased numbers of RPAs have brought a corresponding demand in the volume and types of missions. It has been suggested that UASs may be a relatively low-cost and effective means of securing the nation's borders. Civilian law enforcement has begun to experiment with small surveillance UASs. American warships off the lawless coast of Somali began using them to locate pirates threatening one of the world's most important shipping lanes. In February 2009, unmanned aircraft helped locate and apprehend nine pirates in a skiff after they fired an RPG at a merchant vessel.

As a pilot of the future *now*, I didn't have to worry about the physiological stresses of high-speed flight in my grounded zero-knot, one-G cockpit. I could focus on the tactical situation while backed up by a team of experts spanning the globe. War fighters from the president of the United States down through the Chiefs of Staff and the Pentagon to the grunt hiding behind a mud wall in Baghdad understood that Predator and its kin were truly in the business of saving American lives wherever they flew.